THE BAOFENG RADIO SURVIVAL GUIDE

THE ULTIMATE RESOURCE TO MASTER YOUR RADIO IN EMERGENCIES, DISASTERS, AND GUERRILLA TACTICS & ENHANCE COMMUNICATION SKILLS IN REAL-LIFE SCENARIOS

ZEPH GUERRERO

TABLE OF CONTENTS

INTRODUCTION

Welcome to The Baofeng Radio Survival Guide, the definitive resource for mastering your Baofeng radio and using it to enhance your communication skills in real-life survival scenarios. Within these pages, you will learn everything you need to know to utilize your Baofeng radio effectively in emergencies, disasters, and beyond.

What is a Baofeng radio, and why are they so popular for survival situations? Baofeng is a brand of affordable and programmable handheld two-way radios that operate on FRS, GMRS, MURS, and Ham radio frequencies. Their low cost and wide range of features have made them a popular choice for preppers, hikers, explorers, and those interested in off-grid communication. The UV-5R model, in particular, strikes a great balance between price, functionality, and customizability, earning it the nickname "the Swiss Army knife of radios." With proper use and understanding, a Baofeng radio can become your most valuable tool for maintaining contact in difficult times when cell towers are down. They allow instant, short-range voice communication without any infrastructure requirements or monthly bills. Their true power lies not in any single feature but in their versatility across various emergency contexts, from natural disasters to tactical operations. Within this book, you will learn how to optimize your radio and develop the essential skills to leverage its full potential.

We will start with an overview of Baofeng radio models and a more in-depth focus on the UV-5R. You will gain a technical understanding of its hardware components like the battery, antenna, and controls, as well as the wide range of functions it can perform through user-programmable settings. We will cover how to program channels, configure privacy codes, and optimize other settings using the CHIRP software for Windows PCs. A solid grasp of your radio's technical specifications and capabilities will serve as a strong foundation. From there, we delve into fundamental radio communication concepts. You will learn terminology, standard phonetic alphabet, protocol etiquette, and techniques for effective voice transmissions. Proper radio discipline and following communication best practices are essential for avoiding confusion and maintaining calm, productive contact during stressful incidents. We explore seven optimized strategies for utilizing a two-way radio to solve problems and coordinate responses.

Building on the basics, we explore advanced functions like repeater use, digital modes, and encryption options. Tactical communication requires knowing when and how to employ such tools. Preparedness for any emergency also demands thorough planning. We offer guidelines for developing robust radio systems, standard practices, and prudent battery/power redundancies. Realistic checklists, emergency procedures, and tips for off-grid operation with your Baofeng will prepare you to face life's unpredictable difficulties with greater resilience. Communicating takes on new importance for military, law enforcement, or civilian resistance groups. We cover vital tactical communication principles, encryption methods, and how to establish intuitive plans for coordinated operations. Well-planned verbal coordination can mean the difference between mission success and failure in dangerous environments. Several actionable case studies demonstrate how radios like the Baofeng have been effectively applied across different real-world crises.

Of course, even the hardiest gear requires regular maintenance to perform at its best. We include a troubleshooting section covering the 10 most common radio issues and how to resolve them. Recommendations for useful radio accessories aim to optimize your setup further. Finally, the book concludes with an overview of relevant FCC regulations and licensing requirements to ensure lawful and informed use of public radio frequencies. Whether facing natural disasters, social unrest, or personal emergencies off the grid, a Baofeng radio in trained hands becomes a force multiplier. Within these pages lies the knowledge to develop your skills to a level granting far greater communication abilities and peace of mind in adversity. I hope you find this comprehensive book as invaluable an ally as the radio itself. May it help you prepare to overcome life's hardships with greater strength, resources, and community. Welcome to the world of two-way radio survival communication - let's get started!

CHAPTER 1

Introduction to Baofeng Radio

About the Radio

Baofeng radios are affordable two-way handheld radios that have gained widespread popularity in a variety of hobbies and use cases due to their combination of performance and value. This section will delve into the technical details and capabilities of Baofeng radios so users have a thorough understanding of how they work.

What is a Handheld Radio?

Handheld radios are portable, self-contained, two-way radio transceivers that are powered through internal rechargeable battery packs. They are designed to be easily carried and allow users to communicate via push-to-talk functions. Various public safety agencies, businesses, and hobbyists have used handheld radios for decades to maintain contact over short-to-medium distances depending on terrain and conditions. Baofengs are a type of handheld transceiver made by the Chinese company BaoFeng Tech. While their build quality is sometimes criticized for being cheaper than higher-end brands, they have carved out a devoted customer base thanks to their impressively low price. It's important to note that despite their affordable cost, Baofengs are fully-featured digital radios capable of operating on a wide array of frequencies when properly programmed.

Radio Frequency Fundamentals

 All radios transmit and receive electromagnetic radio waves, which are variations in electric and magnetic fields that propagate at the speed of light. Radio waves are classified based on their wavelength, which is inversely proportional to frequency - higher frequencies have shorter wavelengths and travel farther than those with longer wavelengths given the same transmitting power. Radio communication is possible when two radios are tuned to the same frequency. The transmitting radio varies its radio frequency carrier wave by imposing audio signals from the microphone onto it. This modulated signal propagates through space and can be picked up by another radio tuned to that frequency, which then extracts the audio information and plays it through the speaker. Different radio systems work on separate portions of the overall radio frequency spectrum as allocated by regulatory bodies.

Baofeng Frequency Capabilities

Baofeng radios, like the popular UV-5R, are often described as dual-band because they can transmit and receive on both VHF and UHF frequency bands. This allows you to access different radio services:

- VHF High Band (136-174 MHz): Covers FRS/GMRS (Family Radio Service/General Mobile Radio Service), MURS (Multi-Use Radio Service), as well as amateur "ham" radio 2-meter band.
- UHF Band (400-480 MHz): Includes 70cm amateur band as well as commercial radio systems like P25 digital trunking networks used by many public safety agencies.

In addition, the receiver on Baofengs can usually tune down into the VHF low band from 30-88 MHz to pick up broadcasts and unmodulated signals. This wide range makes them very versatile scanners even if transmitting abilities are limited in certain sub-bands. It should be noted that FCC rules restrict transmitter power levels and frequencies Baofengs can legally operate on depending on the radio license held.

Inside the Baofeng Radio

Let's explore the key internal components that allow Baofengs to function:

CPU: An onboard microcontroller runs the radio's full digital signal processing and user interface. This provides features like configurable channels, tone alerts, and menu settings.

Transmitter/Receiver Circuits: Baofengs contain both transmitter and receiver radio frequency circuits to generate signals for transmission and amplify weak signals for reception. Common module types include single-conversion and dual-conversion superheterodyne receivers.

Controller Buttons: Programmable buttons allow direct access to channels/zones as well as menu options for settings configuration when held. Side buttons select emission type (FM/AM/narrowband FM) and power/volume.

Display Screen: Liquid crystal displays show the readout of frequency or channel numbers, transmitter power levels, settings menus, and more through an intuitive interface. Backlighting aids visibility.

External Antenna Port: SMA male connectors (usually reverse SMA on mobile units) attach detachable antenna solutions suited to the environment and frequency range in use. On-board "rubber duck" antennas provide basic performance.

Internal Battery & Charging: Rechargeable lithium-ion polymer cells in common capacities of 800-2600mAh power each radio. USB ports allow recharging without removing the battery pack.

Options Port: Additional connection sold separately facilitates direct computer programming, external speaker-mics, remote switch hookups, and more advanced features.

Selecting Communication Modes

Depending on application and licensing restrictions, Baofengs offer programming options for several transmission modulation types:

FM Narrowband (12.5 kHz): The most common mode provides efficient channel spacing for licensed services like FRS/GMRS and ham radio. Lower noise and clearer audio than wideband FM.

FM Wideband (25 kHz): For short-range line-of-sight use on license-free frequencies. Offers improved range but lower voice quality and reduced channel capacity versus narrowband.

AM: Only relevant for ham bands below 30 MHz where voice modes use Amplitude Modulation. Ranges are less than FM, but AM waves diffract around obstacles better for longer-distance communications.

DMR Digital: Some higher-end radios provide an option for DMR digital transmission/reception. Compared to analog FM, DMR delivers a longer range through TDMA channel sharing and higher voice quality through more efficient encoding.

Understanding Channels, Groups & Signaling

As programmable digital radios, Baofengs offer versatile channel maps to transmit and receive across associated frequencies. Key organizational concepts include:

Channels: Individual radio frequency assignments are shown numerically on the display. Programmed into radios to provide quick access to repeater systems or simplex calling frequencies.

Groups: Logical sub-channel designations used on ham/FRS systems and P25 trunked networks. Unique squelch/alert tones select communications from within groups on shared repeaters or wide-coverage systems without manually tuning.

Tones: Sub-audible tones transmitted before/during transmissions to access certain repeaters and open squelch on receiving radios in the same group. It is required for P25 digital mode signaling as well.

Proper configuration of channels, groups, and tones based on intended radio networks ensures reliable selective calling and access to diverse communications systems supported by Baofengs. This improves real-world functionality versus direct frequency selection alone.

Applying Accessories and Upgrades

The modular construction of Baofeng radios supports a thriving third-party market for compatible accessories aimed at boosting performance. Some key aftermarket options include:

Antennas: Replace stock rubber antennas with better range adaptations for VHF, UHF, or dual-band from options like Nagoya or Diamond antennas.

Speaker-Microphones: Add convenient push-to-talk capability and improved ergonomics beyond the standard controls. Some integrate features like illuminated buttons.

Batteries: Larger-capacity packs extend operating time away from power between charges. Quick-disconnect designs aid hot-swapping without tools.

Programming Cables: Direct PC interfaces simplify firmware updating and advanced digital programming beyond the radio interface.

Car Kits: Mobile mounting brackets and wired mic/speaker bundles configure handhelds for fixed vehicle operation. Some include external antenna ports.

Earpieces: Discrete communications discreet earpiece attachments maintain inconspicuous monitoring without disturbing others.

With thoughtful consideration of intended applications and channels of use, Baofengs and their accessories enable highly customized radio solutions far exceeding basic performance at very reasonable budgets. Their modular open design fosters ongoing innovation by third parties as well.

Baofeng Model List

As one of the leading manufacturers of affordable radios, Baofeng Tech offers a wide selection of models suited to various applications, frequencies, and customer preferences. This section provides an overview of some of the most popular handheld transceiver options, along with their key specifications and use scenarios. While the low-cost Baofengs open doors to practical radio communication, choosing the right radio depends on intended operation parameters.

UV-5R

Considered by many as the best all-around entry-level model, the UV-5R remains a top-selling Baofeng to this day. Covering 136-174 MHz VHF and 400-480 MHz UHF bands with 8-watt RF output, it offers a versatile foundation for local talk groups and monitoring ham and public service frequencies flexibly. Dual watch, 121 memory channels, and CTCSS/DCS signaling provide basic trunking and repeater utilities. At less than $30 retail, its value is tremendous.

UV-82HP

A direct successor building upon the UV-5R's formula, the UV-82HP upgrades to 16 memory channels, a high/low power switch, and removable antenna mounting for better connectivity. Keeping the same wide frequency range, it also adds new features such as VOX voice-activation of transmit and compact keypad buttons for easy operation with gloves or wet hands. Battery life sees improvements as well, making this budget model better suited to dispatcher duties and emergency applications.

GT-3WP

For mobile installations, the Baofeng GT-3WP brings handheld functionality to vehicle environments using external mic/speaker bundles. Its cigarette lighter power cable charges an internal pack while its control head sits neatly on car/ boat dashboards. Benefiting from higher authorized power at 50W, the GT-3WP can expand coverage of business repeaters even while mobile. Memory banks are proportionately larger as well at 32 channels to handle diverse locations and rotations efficiently.

UV-5X3

As a step up in capabilities, the UV-5X3 provides premium features at an affordable mid-range cost. Dual watch on independent VFO frequencies, expanded memory capacity to 128 channels, and 400/800mW switchable transmit power offer added flexibility over basic models. Its models. Its top-facing control button design also feels more robust for heavy daily use in field applications like security, logistics, or utility worksites where communications are vital. Control head removal further aids mobile mounting.

UV-82X3

Offering an even more advanced digital experience on a budget, the UV-82X3 is one of Baofeng's flagships for serious amateur or public service usage within FCC rules. It operates on multiple air interfaces, including DMR Tier I (less than 1W output power) as well as conventional analog FM. Features like dual standby Tier I/analog scanning and Bluetooth connectivity for programming set it apart from purely analog peers. Among available models, it provides the most cutting-edge digital functions without a high-dollar price tag upfront.

UV-9X3

As a more powerful flagship set, the UV-9X3 raises specifications across the board for intensive critical infrastructure or emergency response situations. In addition to DMR and 800/4000mW power settings, it supports both digital AND analog transmission simultaneously on separate frequencies via dualwatch. 200-channel memory, GPS navigation interface, and removable high-capacity battery packs make this a true portable communications hub. For professionals working in hazardous duty areas, its utility comes at a remarkably low total cost of ownership.

UV-5XPRO

Marketed towards serious amateur radio operators looking for budget breaker functionality, the UV-5XPRO delivers tiered specifications beyond typical FRS/ GMRS applications. Differences include a removable 7.4V 1800mAh battery with a fast heating charger, dual PTT buttons (one digitally retained), 121 privacy codes, VOX voice-activation of transmit, GPS coordinates queries, and more parameters configurable via menus. Though targeting sophisticated ham features, its price tag remains within reach of value-minded enthusiasts or backup radios for clubs.

Choosing Based on Needs

While all Baofeng models share affordable pricing beneficial to experimenters and frugal users alike, evaluating key factors helps select the right product match for circumstances. Factors like frequency bands, UI simplicity, analog vs digital modes, accessory eco-systems, and power requirements steer choices towards optimized solutions, whether personal, industrial, or public safety focused. Understanding the purpose and value propositions behind the varied brand families enables radio selection to fit intended operational needs at every step of the operating scale.

Accessories Boosters

Maximizing returns on any Baofeng investment also means leveraging its modular platform compatibility to expand performance through affordable accessories. Common, cost-effective upgrades include extended battery packs for events/field work, stubby antennas boosting localized connectivity without extra size, remote speaker-mics adding hands-free flexibility in vehicles or vehicles, and programming cables optimizing digital functions through software. Interchangeable infrastructure expands utility room utility over lifecycles in a way proprietary offerings resist. For inspired operators, the potential for continual upgrades over time enhances value well beyond initial purchase costs.

Focus on UV-5R: The Best Model in The US

Among the various models offered by Baofeng, the UV-5R stands above the rest as the single best option for radio users in the United States. Its combination of affordability, versatile capabilities, and strong aftermarket community make it uniquely well-suited to amateur radio, public service scanning, and general survival communications needs within FCC guidelines.

Unbeatable Price Point

Perhaps the strongest selling point of the UV-5R is its remarkably low cost of entry. Priced consistently around $30-40 dollars depending on included accessories, it undercuts all but the most basic bubble-pack FRS radios while outperforming them in practically every way. For hobbyists just starting out, prepper groups building collaborative response networks, or individuals seeking pragmatic self-reliance tools, few pieces of communications gear can match the UV-5R's value. Its affordable nature brings practical hands-on radio experience within financial reach for all walks of life.

Dual-Band Frequency Coverage

As a dual-band radio, the UV-5R easily spans the most relevant and license-exempt public service, business, and amateur frequencies used across the U.S. Its 136-174 MHz VHF coverage permits both 2-meter ham use and monitoring of trunked public safety systems in the high VHF band. The 400-480 MHz UHF range similarly captures 70cm ham action plus any 400 MHz commercial entities like Forestry Service in a given locale. Whether for transmitting simplex/repeater calls or just scanning interesting channels of activity, this breadth of supported spectra maximizes potential utility right out of the box.

Wide Receiver Capability

While its frequency synthesis allows transmitting only within narrow authorized bands, the UV-5R's receiver persists as an impressive asset for radio surveillance and situational awareness. Offering a full scanner range from 66-88 MHz low band VHF up through 1 GHz UHF/SHF, any signals or broadcasts throughout this wide swath may be listened in on. From traditional 2m/70cm amateur voice to FM/TV broadcasts and even aircraft telemetry in places, the receiver component of the UV-5R serves well as a small and portable communication intelligence gathering tool.

Deep Aftermarket Support

Given its immense popularity within amateur circles and preparedness communities, third-party accessory companies have fully embraced the UV-5R with a massive range of compatible upgrades and additions. From higher-capacity battery solutions and wider-bandwidth antennas to programming cables, microphones, and even APRS digital packet capabilities, tinker-friendly owners can continually refine their UV-5R systems through affordable modular pieces. A self-sustaining upgrade ecosystem striving to better the platform keeps its relevance and performance potential growing over the long-term through grassroots innovation. Whether for simple field repairs or custom functionality, no other radio matches this level of supplemental support.

Programmable Memory Banks

At the heart of the UV-5R's utility lies its ability to store over 100 customizable memory channels, mapping frequency locations and signaling parameters for diverse radio systems. From simplex communications plans to complex trunked networks, structured channel maps accessible via 12.5 kHz step scanning or direct channel selection facilitate seamless repeated access via pre-programmed profiles. Its addressable memory system lays the logical groundwork for multiple operational configurations, from personal to large-scale coordinated response applications managed through centralized channels.

Long-Term Value in a Harsh World

When evaluating tools for surviving severe disruptions outside normal services and infrastructure, a radio's ability to connect people across distances proves invaluable for matters of safety, navigation assistance, or just basic human contact during hard times. The UV-5R offers this capacity at a price so affordable that even those on limited budgets may consider several units for a group's cooperative readiness. Paired with take-along essentials like food, protection, and first aid, owning a UV-5R radiates prepared confidence that help and calm voices may reach out even in darkness. Through collaboration over the airwaves, communities strengthen their collective chances. Such an easily accessible support system, uniform across economic brackets, is a stabilizing societal safeguard to embrace proactively for potential crises ahead.

Understanding the Power of Baofeng Radios in Survival Situations

While Baofeng radios offer an affordable gateway to the exciting world of radio communications as a hobby, their true potential lies in how their low cost and portable capabilities enable people to stay connected through emergencies and survive hard times. This section explores the valuable role handheld transceivers can play in preparedness when disruptions to normal infrastructure occur.

Lifelines for Coordination

Whether due to disasters, social upheaval, or simple geographic isolation, loss of cell networks and limited landlines leaves people and groups cut off without practical means to coordinate efforts or request aid. Radios prevent this disconnect by maintaining a wireless backchannel for basic telephony-style voice calls independent of external services. With organized plans, radios can effectively network neighborhoods into command structures, facilitating everything from search and rescue missions to supply deliveries during crises when every minute counts.

Mobile Versatility

Unlike fixed-location shortwave broadcasts vulnerable to interference, the portability of Baofeng radios allows mobile communications on-the-go through changing conditions. Hikers, for example, stay in contact with staging areas even far from roads or buildings as they partner for missions ranging from evacuations to resource collection. First responders leverage this mobility across unpredictable incidents from fires to floods, guiding assets efficiently while addressing dangers. For evacuees on the move, radios provide reassuring coordination, improving chances.

Situational Awareness

Radios transform isolated individuals into nodes within an open-air sensor network, crowd-sourcing observations that paint a comprehensive real-time picture of unfolding threats across wide scopes. Reports of dangers like downed powerlines, washed-out bridges, and roaming criminal elements flow wirelessly to offer warning and aid tracking of evolving dangers. Combined updates develop a shared intelligence to maneuver around or prepare countermeasures as a utility against unknown risks across infrastructure breakdowns.

Affordable Redundancy

Where expensive commercial land mobile radio systems carried by agencies remain centralized chokepoints, the low cost of Baofeng ownership enables mass-distributed capabilities. Individuals act as voluntary "auxiliary communicators" using their privately owned gear. This decentralized, organically sustainable model forms independent backup skeletons, maintaining a safety net should larger centralized systems fail due to lack of power, damage, or signal jamming sabotage. Redundancy increases the chances of getting through no matter what.

Mental Health Support

Isolation breeds anxiety, fatigue, and discord, which are detrimental to survival. Radios lift psychological burdens through routine check-ins, status updates, friendly exchanges, and even entertainment/news broadcasts reminding of the larger connected world beyond immediate circumstances. Consistent social contact alleviates stresses that jeopardize judgment and group cohesion over long emergency separations. For some, the calm voices may prove a literal lifeline preventing discouragement's descent into the darkest outcomes.

Extending Communications Ranges

Through repeaters, which rebroadcast signals to extend their distances, or simple relay stations set up opportunistically, radio signals can span further than line-of-sight limitations normally allow. This multiplies each transmission's useful coverage, weaving a resilient web of radio-mesh connectivity even in cut-off hamlets, valleys, or wrecked city blocks where cell/internet black holes would otherwise leave people lost to reach help. Distance collapse brings hope.

Learning & Improving Readiness

Even outside emergencies, radios spark curiosity and science while strengthening preparedness, knowledge, and bonds between neighbors. Hobby radio operators experiment with antennas, propagation, codes, and technical skills transferable to survival scenarios, gaining valuable familiarity before the need arises. Preparedness groups organize practical drills applying lessons learned. Regular airwave nets build camaraderie and share stored wisdom, renewing a resolve to look out for one another through any trouble ahead.

CHAPTER 02

Features and Functions

What can you do with a Baofeng radio UV-5R

The Baofeng UV-5R is an affordable multi-purpose handheld radio that can be used for various amateur radio activities and emergency communications purposes. While basic out of the box, it has capabilities that allow owners to explore many uses once configured properly.

Listening

One of the first things new radio owners enjoy is listening to what's happening on different frequencies. Right out of the box, the UV-5R can receive signals across multiple frequency bands, including VHF and UHF ranges. This allows listeners to monitor public safety communications, weather radio stations, ham radio conversations, FRS/GMRS channels, airband, and railroad communications. The large frequency range and rotary tuner knob make it easy to scan or search for interesting signals to familiarize themselves with radio operations and procedures in different services.

Two-Way Communications

Once programmed with frequencies and tones if needed, the UV-5R allows licensed amateur radio operators to conduct two-way voice conversations using technologies like analog FM and DMR digital modes. Common uses include calling CQ on local repeaters to meet other hams, checking into weekly radio nets hosted on repeaters to receive news and bulletins, portable operations during hiking, boating, or camping trips using simplex frequencies, participating in public service events and drills by communicating with the command post, emergency communications by linking up with local ARES/RACES teams for storms or disasters and linked repeater networks for longer range simplex communications using digital mode and tone signs. The 5-watt output power and removable antenna allow effective mobile and portable use within typical repeater coverage areas.

Data Modes

While best known as a voice radio, the UV-5R is also capable of basic digital data modes with proper setup, including APRS, PSK31, DTMF, DMR, and FSQ. This opens up additional digital communication options beyond voice, including the transmission of weather reports, bulletins, telemetry sensors, and more. A USB cable allows computer-based data modes when combined with free software programs.

Programming and Customization

Out-of-the-box operation covers the basics, but programming unlocks the full potential of the UV-5R radio, including programming frequency offsets, creating memory bank channels, setting lockout options, programming DCS/DCS codes, setting transmit power levels and tuning steps, configuring display features, programming DMR talk groups and timeslots, and installing optional firmware updates. CHIRP or direct programming allows tailoring the radio specifically for a user's local area, needs, and hobby interests.

Optional Accessories

A number of aftermarket accessories can extend the abilities of the UV-5R, such as dual-band antennas, speaker mics, remote speaker mics, programming cables, Li-ion battery packs, signal playback devices, mounting brackets, earpieces, and more.

Emergency Applications

When properly prepared, the UV-5R provides an affordable communication solution for basic emergency and disaster response needs, including linking up with ARES/RACES emergency coordination nets, serving as a backup radio, maintaining contact between utility/relief workers, coordinating medical evacuation requests, passing real-time weather/evacuation updates and more.

Programming the Baofeng UV-5R: A Step-by-Step Guide

Out of the box, the Baofeng UV-5R radio can operate on default settings; it needs to be programmed with your specific frequencies, channels, and settings. This guide will walk through the process of programming the UV-5R manually via its front panel controls in a step-by-step manner.

Equipment Needed

- Baofeng UV-5R radio
- List of frequencies and settings to program
- Programming manual/chart for reference

Programming Mode

To access the programming mode on the UV-5R:

1. Turn the radio off
2. Press and hold the leftmost (#) key while turning the radio on
3. "GMT" will appear on the display, indicating that the programming mode is active

Navigating Menus

Use the up/down arrow keys to scroll through menu options and the left/right arrow keys to select sub-menus or toggle settings. Press PTT to confirm selections.

Setting Basic Channels

1. Use arrow keys to select "Channel Setup," then press PTT
2. Select an empty channel to program using left/right arrows
3. Press PTT to edit, and "MR" will flash

4. Use arrow keys to input frequency (include decimal for MHz)
5. Press PTT to save frequency and exit to the previous menu

Adding Channel Names

1. Select a programmed channel
2. Press PTT, "MR," and the first letter will flash
3. Use arrow keys to input channel name (max of 8 characters)
4. Press PTT to exit the name input

Setting Channel Groups

1. Select "Group Setup" and press PTT
2. Use arrows to select the Empty group
3. Press PTT to edit, and "GP" will flash
4. Use arrows to select channels for that group
5. Press PTT to save a group

Customizing Settings

Use the same process; additional settings can be customized under menus like Preference Settings, Scan List, and more according to your needs.

finishing Programming

1. Once all desired channels and settings are input, select "Programming Complete."
2. Press PTT and "Complete" to save changes
3. Turn the radio off to exit programming mode

Saved frequencies and settings can be copied to other radios using the cloning feature to synchronize radios in the field without reprogramming from scratch each time. Familiarize yourself with the newly programmed channels by navigating with the channel selector buttons before use in the field. Practice accessing channels by name using the rotary tuner as well. Have fun exploring your new radio!

If errors occur, double-check all values against the programming sheet. If issues persist, try resetting to factory defaults and reprogramming from the start.

Programming with CHIRP Software

CHIRP is a free and open-source programming software that allows the configuration of radios through a computer interface. Compared to manual front panel programming, CHIRP streamlines the process and enables advanced features. This guide will outline how to program the Baofeng UV-5R using CHIRP software on Windows.

Equipment Needed

· UV-5R radio
· USB data cable
· Computer with CHIRP installed
· Frequency/setting programming file

Downloading & Installing CHIRP

1. Visit www.chirp.danplanet.com and download the latest version of CHIRP for Windows.
2. Run the installer and follow the on-screen prompts to complete the installation.
3. Optionally install driver files if programming other radio models

Connecting the Radio

1. Turn the radio off and connect the USB data cable
2. CHIRP will auto-detect supported radios once the cable is plugged in
3. If not found, drivers may need to be installed separately
4. The radio appears in the "Radio" dropdown list once it is detected

Creating a New File

1. In CHIRP, click "File" then "New" to generate a blank file
2. Select UV-5R from the "Radio" list
3. Click "OK" to save the empty file

Programming VHF/UHF Frequencies

1. Click the "Memory" tab and right-click an empty channel
2. Select "Edit," then enter frequency, offset if needed
3. Click "OK" to save that memory
4. Repeat for other desired frequencies

Adding Channel Names

1. Right-click the programmed memory
2. Select "Edit," then enter a name up to 8 characters
3. Adjust the "Label" column width for visibility
4. Click "OK" when finished naming

Configuring Advanced Settings

Additional sections like "Settings," "Tones," and "DTCS" allow setting things like transmit power, CTCSS tones, DCS codes, and more as needed. Options will vary between radios.

Transferring to the Radio

1. Ensure the radio is securely connected via USB
2. Click the "Read" button to retrieve current radio settings
3. Click "Write" to send the new file contents to the radio
4. A progress bar confirms the process is underway

Opening/Editing Existing Files

Previously saved CHIRP files can be accessed by clicking "File" and then "Open." Any memories or settings can then be modified as needed before sending an update to the radio.

Printing a Programming Sheet

For documenting the programmed contents, click "File" then "Print." Adjust layout and paper settings before sending to a printer for a physical programming reference sheet.

Programming Multiple Radios

Finished files can be cloned to other radios of the same model by writing the file while those radios are connected one after the other. This synchronizes them without reprogramming each individually.

Tips for Best Results

1. Ensure the latest driver/firmware is being used
2. Use a high-quality USB data cable
3. Close all other apps and minimize wireless interference

4. Double-check all values before sending them to the radio
5. Troubleshooting Connection Issues

If the radio isn't detected, try a different USB cable or USB port or restart the computer and radio. Drivers may need to be manually installed as well. CHIRP support forums can assist with any persistent communication errors.

Optimized Settings & Customization

Out-of-the-box settings on the Baofeng UV-5R are practical for basic use, but more advanced owners can tweak preferences to better suit their radio environment and intended applications. Understanding the adjustment of configuration options also facilitates keeping backups and reset options.

Transmit Power Level

The default power output is high at 5 watts. This can be adjusted down for portable use or where regulations require. Lower settings improve battery life at the expense of range.

To change in programming mode:

1. Select "Preference Setup"
2. Scroll to "Transmit Power" and press PTT
3. Use arrows to choose from high/Mid/Low settings

Frequency Tuning Step

The default 25kHz step works well for most VHF/UHF, but a narrower tuning increment can be preferable.

To modify tuning step size:

1. Select "Preference Setup"
2. Choose "Frequency Step" and press PTT
3. Arrow down to 5/6.25/10kHz option as preferred

Display Illumination

The LCD backlight can distract or consume the battery rapidly at full brightness. Adjust to balance visibility vs power.

To customize screen brightness:

1. Access "Preference Setup"
2. Select "Luminance" and press PTT
3. Cycle between High/Med/Low options

Key Beep Volume

Some find the button tones bothersome. Adjust the volume without disabling it entirely. Access the same steps as Display Illumination, but select "Beep Level" instead.

Battery Saver

By default, the UV-5R shuts off after 10 minutes of transmissions. Extend the auto-power down timing while portable operating to conserve batteries. Select "Battery Saver" then "Timer" to choose between Normal/Extended options.

VFO/Memory Modes

Experiment with various scan/search settings to become familiar with frequency navigating methods. Enable/disable scan resume to suit preferences. Select the "Scan" sub-menus and options as needed for customizing functionality.

DTCS/DCS Code Settings

Private lines accessible via tone squelch require matching digital privacy codes. Save different sets of codes corresponding to repeaters for quick switching. Navigate to "Privacy Encoder/Decoder" to input desired tone pairs.

DTMF Autodial

Program sequences of automatic DTMF digits for functions like repeater control using the keypad method. Great for opening linked systems. Access DTMF autodial functions under "DTMF Autodial."

PTT ID

Some DMR systems support transmitting a short identifier after pressing PTT, such as a callsign. Enable/disable this feature according to need.

Configure under the "PTT ID" sub-menus.

Display Options

Personalize the UI by modifying colors/styles or turning off non-essential items like channel names/frequencies to reduce screen clutter. Navigate to "Display Setup" and features within for customization.

Radio ID

The unique serial number is included for license tracking on ham bands. Adjust letters/numbers as preferred. Select "Radio ID" for modification.

Cloning Settings

Save cloning profile configurations to easily duplicate current settings to other identical radios with a single button press for on-site rapid synchronization. Access cloning profiles and options under "Clone."

Programming Reset

In the event of issues, return to default factory settings before attempting to reprogram from scratch. Leave a copy of customized programming offline as a backup. Soft/Full resets are accessible in the "Preference Setup" sub-menus.

Antenna Configuration and Usage

The antenna is a critical component impacting radio performance. This chapter explores configuring different antenna types and optimizing usage to extract maximum effectiveness from the Baofeng UV-5R. The focus will be on the selection, installation, and testing of privacy/encryption features.

Antenna Categories

Understand the three main antenna varieties used:

Stock Rubber Ducky: Compact but has low efficiency. Acts as a starting point.

Magnetic Mount (Diamond, Nagoya): Attach to vehicles/metal. It's a better range than stock.

Separate Antenna with Coaxial Cable: Highest gain when remotely mounted higher for greatest coverage.

Selection Factors

Consider antenna specifications like frequency range, power rating, length, and placement location during evaluation for a specific radio's needs. Higher gain designs elevate coverage area.

Installation

Magnetic mount: Clean metal surface and center antenna.

Separate antenna: Attach the PL-259 connector to the radio jack securely, then secure the antenna mount location.

Avoid placing near obstructions or other antennas for efficiency.

Testing Receiving Sensitivity

Monitor local frequencies with different antennas and locations. Strong signals ensure reliable reception at the furthest possible points within coverage parameters. Noting differences helps gauge upgrade effectiveness.

Setting CTCSS/DCS Tones

Private lines require matching subaudible tones/digital codes for access. Program repeaters' tones under Tone/Code Setup so radios only decode signals with correct info for privacy.

Turning On Encoder/Decoder

To use private lines:

1. Access Tone/Code Setup and select Tone/DCS options.
2. Set encoder/decoder to /transmit tones/codes of repeaters.
3. Toggle On in programming.

Testing Transmitting Effectiveness

Contact another station with the same antenna/tone configurations to test connectivity at maximum potential distances. Adjust antennas' orientations as needed for optimal line of sight. Confirm receipt of transmissions.

Encrypted Communications

Some digital modes facilitate encrypted voice/text using pre-programmed encryption keys.

Requirements:

· Radios use the same encryption type (e.g., DMR)
· Identical encryption keys installed
· Both radios' encryption enabled

Test by exchanging private messages undecipherable without a key. Careful setup ensures security.

Additional Privacy Tactics

Other techniques like shortened transmitter duration, altering microphone audio properties, and switching frequencies periodically enhance covertness to prevent easy monitoring. Combining strategies fortifies signal obscurity.

Optimizing Portable Usage

For hiking/contests, stow the radio snugly in a backpack to prevent damage, then use a handheld antenna or attach a magnetic mount externally near the top for ideal reception angles. Adjust the volume so as not to disturb nature.

Vehicle Installations

Secure radio in stable locations away from emissions sources. Hardwire external antennas for reliable permanent mounts with ideal reception. Fuse lines appropriately

CHAPTER 03

Basic Radio Communication

Exploring Communication Fundamental Concepts, Terminology & Channels

Communication Fundamental Concepts

Effective communication is essential when using a radio to coordinate with others. Some key fundamental concepts to understand include clarity, brevity, precision, and protocol. Taking the time to comprehend these principles helps ensure radio communications are understood by all parties involved. Clarity in communication relates to speaking in a clear and easily understandable manner. Shouting into the radio or using unclear jargon and shorthand only introduces the potential for miscommunication. Stress can cause someone to speak too fast or loudly, so maintaining a moderate, steady tone is important for clarity. Brevity means being concise with messages and avoiding unnecessary words or details. Limited radio channel capacity requires keeping transmissions brief so the frequency remains available. Only the essential information needed by the recipient should be transmitted. Everything else wastes time and risks losing important traffic in the channel. Precision involves being exact and unambiguous. Terms, numbers, locations, and other key details must be expressed without room for misinterpretation. Saying "building five" instead of just "five" leaves no question of the message context.

Terminology & Channels

Agreed terminology provides the common language and shared understanding necessary for precise communications between radios. Standard signs, codes, call signs, procedural words, and other terms get everyone literally on the same page when coordinating efforts. Procedural words or prowords are brief, clearly understood intent, and prompt response. Common prowords include:

- Affirmative - Yes, the transmission was received and understood.
- Roger - I have received all of your last transmissions.
- Negative - No, or permission not granted, or that is not correct.
- Say again - Repeat your last transmission.
- Spell - Spell the word or group of words phonetically on the radio.
- Stand by - I must pause for a few moments.
- Correction - An error has been made in the last transmission; the correct version follows.
- Clear - Consider this radio frequency clear of traffic and users.

The phonetic alphabet standardizes spelling letters over the radio to avoid potential confusion between similar-sounding characters. The NATO phonetic alphabet commonly used is:

Alpha, Bravo, Charlie, Delta, Echo, Foxtrot, Golf, Hotel, India, Juliett, Kilo, Lima, Mike, November, Oscar, Papa, Quebec, Romeo, Sierra, Tango, Uniform, Victor, Whiskey, X-ray, Yankee, Zulu.

Additional common radio terms include:

- Radio check - A brief test transmission to check radio signal strength and reception quality.
- Come in, please - This is a request for a response from the station being called.
- Over - I have completed my transmission and am awaiting your response, reply, or traffic.

- Out - This denotes the end of complete contact with an acknowledgment or response expected.
- You are clear - At the end of a contact, indicate the channel is again available for others after the current transmission concludes.
- Some key terms relate to requesting status updates or relaying information:
- Situation report - A concise summary of the current circumstances being monitored.
- Transmission confirmed - Verify the last radio message from another station was successfully received.
- Relay - Passing information received to another party.

All these terms, prowords, and standardized elements provide a common framework for organized, coordinated communications free of ambiguity. Their consistent use fosters clear understanding even in stressful situations.

Etiquette & Efficient Operation

Beyond strict terminology, agreed etiquette guidelines smooth radio interactions. Brief, targeted transmissions with proper identifiers yield the highest channel efficiency. Operators should identify by call sign at the beginning and end of messages along with the intended recipient. Standard format such as situation-message-request increases organization. Acknowledging message receipt prevents wasted retransmissions. Phrases like "roger" or "receipt confirmed" let the sender know their traffic arrived successfully. Always acknowledge emergency or priority calls as quickly as possible. Confirm any instructions given over the radio through repetition to verify comprehension by both parties before acting. This prevents critical errors from unclear directives. If unsure, requests for clarification prevent assumption. Monitor a channel briefly before transmitting to identify whether other traffic exists. Do not break into ongoing communication without coordination. Make yourself aware of any operational schedules for the channel to prevent disrupting scheduled events unintentionally. Avoid excessive personal or unnecessary comments over the radio that waste everyone's limited time. Radios should be treated as a professional coordination tool rather than a casual conversation medium.

Golden Rules for Radio Communication

For radios to function effectively during critical incidents, certain guidelines maximize communication clarity, brevity, and cooperation. Adopting these "golden rules" as standard operating procedures helps all operators use the available spectrum productively, even under pressure.

Speak Clearly and Concisely

Stress during a crisis can cause operators to lapse from precise, organized transmissions. To prevent miscommunications, all operators must speak clearly and avoid yelling, fast speaking, or excessive verbiage over the radio. Keep transmissions as brief as possible while still transmitting essential details. Rapid-fire, run-on sentences burden the channel unnecessarily and risk important information getting missed. Take a moment to organize key facts before keying the mic. Speak at a natural, moderate pace to allow understanding without forcing listeners to ask for repetition. Calm, controlled speech maintains clarity and order on shared frequencies even as a scene escalates. Responders worldwide agree clear, concise communication saves lives.

Listen Before Transmitting

Always take a moment to ensure the channel is clear before breaking squelch. Multiple parties transmitting causes disorganization and waste spectrum as intersecting conversations cannot be followed. Listen for several seconds to identify any ongoing traffic that should not be interrupted. Certain transmissions relating to emergency details or safety commands obviously take the highest priority. If a transmission is occurring, wait until its conclusion, denoted by the over/out prowords, before keying up with your own traffic. This allows other traffic to flow smoothly without disruption between operators.

Think Before Transmitting

Refrain from speaking until a message is fully formulated to avoid multiple partial transmissions that clutter the airwaves needlessly. The organization fosters clarity over chaotic conditions. Confirm any traffic includes only the essential details for the intended recipient without superfluous commentary. Other conversations and topics distract from the coordinated mission. Everything broadcast delays other operational communications. Monitor radios regularly to stay aware of developing situations before asking questions already answered. Repeating useful information prevents channel congestion and is better spent coordinating necessary activities. Listen first, so transmissions directly support operations.

Use Recognized Procedures

Consistently employ standard radio protocols, procedures, format, and terminology agreed upon by all agencies on the shared frequencies. Adhering to recognized communications discipline organizes disparate operators into a coordinated response network rather than an anarchic free-for-all. Proper procedures include identifying stations by assigned call signs, acknowledging message receipt, relaying traffic between units, requesting status or situation reports, and so on. Idiosyncratic habits confuse others unfamiliar with non-standard practices.

Confirm Critical Instructions

Verification prevents disastrous errors from ambiguous or miscopied directives that could endanger lives. Repeat back instructions, important call signs, or critical details to obtain confirmation from the sending station before acting on the received traffic. Lack of confirmation leaves uncertainty. As stressful situations escalate, hearing and copying accuracy degrades. Double-checking prevents snap judgments from potential misunderstandings that could compound unfolding problems. Lives saved through such discipline outweigh any minor expenditure of time confirming instructions.

Monitor Radio Status

Check battery levels and radio conditions routinely to avoid dead air failures at inopportune moments. Technical difficulties may require redeployment or interunit workarounds to sustain coordination capabilities. Spare batteries and backup radios allow minimal disruptions. Conduct radio checks periodically to identify any reception or transmission issues that develop -dead zones, interference sources, faulty equipment, and so on before depending entirely on a problematic link during peak operations. Problems addressed proactively pose fewer last-minute difficulties.

Keep Information Secured

Treat all traffic copied as potentially sensitive and not for public exposure, especially operational details, locations, codes/calls. Even friendly operators in adjacent jurisdictions may not require all intercepted traffic. Loose lips threaten safety and compromise prepared responses should situations spiral further. Radio discipline maintains professionalism expected under fire.

Follow the Chain of Command

Coordination depends on designated supervisors coordinating disparate units into a cohesive whole. Operators should route traffic through authorized points of contact and follow supervisory instructions unless obvious safety risks require deviation that can then be explained later. Going rogue breeds dangerous chaos during emergencies, demanding the highest discipline.

Seven Optimized Ways to Use a Two-Way Radio in a Critical Situation

Keep Transmissions Brief

In high-stress incidents, one of the most important rules is to keep radio transmissions as short and concise as possible. Lengthy explanations or discussions squander valuable airtime that could be used to coordinate lifesaving tasks. Only transmit the key details that are necessary for responding units to know, such as critical updates, new instructions, or task assignments. Extra commentary distracts from coordination efforts. Even friendly greetings or casual check-ins tie up the frequency unnecessarily when seconds count. Use recognized message formatting and jargon to efficiently convey maximum useful information in the fewest words. Well, under 30 seconds per transmission should be the goal, allowing other traffic to flow smoothly. Few things sabotage responses more than a monopolized radio channel.

Communicate Only Essentials

Along with brevity, operators must thoughtfully select what merits broadcast over the shared frequency. Not every stray observation or minor report requires an announcement to all users. Transmit only information that impacts immediate response operations or directly involves the safety of personnel on scene. Save non-critical transmissions for less pressing times. Reports must support situational awareness, strategy, resource allocation, or unit tasks to stay pertinent. When circumstances demand repeated or lengthy traffic, designate alternative liaison radio channels or move some communications offline to phones/text if coverage allows. This reserves primary response frequencies solely for coordinated field activities. Unburdening airwaves multiply their utility.

Speak Clearly without Yelling

Stress hormones increase adrenaline, sometimes provoking louder, less controlled speech on the radio. But yelling serves no good - it obscures enunciation and wastes battery life. Speak plainly and firmly without shouting. Maintain a clear, moderate, steady tone, volume, and speed. Express yourself understandably without rushing or mumbling. Enunciate every word distinctly so the message can be copied 100% accurately, even in loud, hectic environments. Perfect communications convey commands, and situation reports faultlessly through noise and interference.

Listen Before Transmitting

Never interrupt active transmissions in progress. Always monitor activity for a few seconds and delay sending your own message until the channel falls clear between outgoing traffic. Barging in mid-conversation fragments communications as competing signals merge into static. Respect that others depend on the airwaves too, so take your turn politely after clear reception indicates floor availability again. Units cannot coordinate when everyone transmits simultaneously into a muddled din. The order stems from listened-before-spoken discipline.

Use Correct Procedures

Coordinated multi-agency efforts require standardized procedures, terminology, and message formats recognized by all participants. Ad-libbed variations confuse other operators unfamiliar with deviations from established norms. Proper procedures integrate all responders seamlessly into the operation through systematic, organized use of the radio spectrum. This entails identifying stations by call signs, using brevity codes if needed, acknowledging messages, phonetically spelling critical details as required, requesting/providing situation updates on schedule, and so forth. Everything about radio communications must function like a well-rehearsed machine supporting incident objectives, not a disjoined free-for-all of mismatched practices between camps. Discipline conquers chaos.

Double-Check Instructions

In high-pressure emergencies, perceptions deteriorate, and instructions prone to mishearing exist. Always have critical directives or tasks confirmed transmission-to-reception by verbally repeating back what you understood for verification. Requesting confirmation prevents catastrophic mistakes from unclear or miscopied traffic that would seriously endanger responders in volatile scenarios. A few seconds of verifying avoids life-risking uncertainties. Discipline applies whether under fire or in normal duties - it forges reliability even through disorder.

Keep the Channel Clear Unless Needed

When not directly involved in traffic, switch to monitor mode to avoid blocking other operations unintentionally. Transmit purposefully with essential communications, then instantly clear the channel upon completion by quick microphone button release. Hogging shared frequencies provides openings for chaos. Responders require maximum coordination capability, so selfishly dominating airtime damages response effectiveness considerably. Selfless adherence from every operator maximizes the spectrum for all. Communication serves the mission above individuals.

CHAPTER 04

Advanced Radio Communication

The Importance of Advanced Radio Communication

A dvanced radio communication systems provide numerous benefits over basic analog radios. Being able to reliably communicate over longer distances, through obstacles, and transmit different types of data beyond just voice can be critical in many situations.

Long-Range Communication

One of the most significant advantages of advanced radio systems is their ability to communicate over longer ranges compared to standard handheld radios.

Basic radios like Baofengs generally have an effective range of only a few miles in open terrain before the signal starts to degrade. While sufficient for small-scale use, this limited range becomes problematic in many real-world scenarios. Advanced technologies allow organizations and groups to communicate reliably over distances up to 100 miles or more, depending on the environment, infrastructure, and equipment used.

Communication through Obstacles

In addition to distance, advanced radio networks can also transmit signals through obstacles that would block standard transmissions. Things like tall buildings and structures, dense foliage, and underground areas frequently attenuate radio waves to the point of losing contact. However, repeaters and digital protocols have enabled reliable communications even in obstructed environments. Emergency responders entering basements, firefighters navigating burned-out buildings, and utility workers inspecting underground infrastructure all depend on radios functioning properly despite intervening materials. Miners, tunnel crews, and cave rescue teams also need off-grid solutions to coordinate inside enclosed subterranean spaces isolated from surface radio coverage.

Transmission of Additional Data

Beyond augmented voice transmission distances, advanced radio systems now afford the capability to exchange different types of critical data over the air. Whereas basic analog radios only support audio transmissions, newer digital technologies enable a host of added information to be relayed.

For example, first responders can transmit photos, building plans, and hazard details from incidents back to commanders for improved situational awareness. Medical personnel can send patient vital signs and histories to awaiting hospital teams. Utility crews can report GPS locations, schematics, and outage details. Sar teams can push missing person descriptions, terrain maps, and the last known coordinates of overdue subjects. Integrating radio systems with technologies like APRS, DMR, and P25 further expands data transmission potential to include preformatted text messages, asset tracking beacons, and even live video streaming.

Systems

In addition to distance, obstacle, and data advantages, advanced radio solutions also deliver more robust system designs, improving the reliability of communications. Things like redundant paths, error correction, interference mitigation, and continuous network monitoring contribute to this greater dependability.

Where basic analog radios rely on localized point-to-point transmissions prone to dropouts from interference, advanced infrastructure takes a multi-node approach. Signals hop through networked repeaters along backup routes if the primary link falters. Digital protocols likewise utilize error correction ECC to detect and request resends of garbled packets. System administrators can also remotely diagnose network health parameters. They proactively address things like weak antennas, intermittent links, or congested repeaters before outages occur rather than reactively troubleshooting after a problem starts impacting users.

Security of Communications

Beyond performance enhancements, advanced systems also apply encryption, authentication, and access controls, enhancing radio security compared to basic analog alternatives. Digital protocols enable privacy features preventing casual eavesdropping as well as restricting transmissions only to validated users on validated devices. For public safety agencies, military units, and industrial operators, transmitting sensitive tactical plans, logistical details, or proprietary information over open analog airwaves risks compromise. Advanced encryption shields discussions, while digital authentication prevents masquerading as another radio. Network administrators can also remotely disable lost or stolen portables.

Future Enhancements

As communications technologies progress, advanced radio benefits will only continue expanding. Emerging areas like software-defined radios enable more flexible protocol support and network augmentations. Integrating radios with IoT systems opens the potential for remote radio control via smart devices as well as incorporating diverse sensors and appliances under a common platform.

Incorporating cellular, WiFi, and satellite modes into robust hybrid PTT networks promises truly ubiquitous off-grid communications. Meanwhile, narrow banding, digital voice coding, and interference cancellation squeeze more utilization from constrained spectrum allocations. Artificial intelligence may eventually autonomously optimize network parameters and even detect anomalies indicating issues requiring attention.

The Installation of Advanced Radio Communication

Establishing advanced radio networks requires careful planning and coordinated installation beyond simply setting up individual radios. Designing a system architecture, acquiring permits, constructing infrastructure, and integrating networking components are just some of the important implementation steps.

System Design

The first phase involves high-level design of the radio network architecture based on operational needs and geographic considerations. Key aspects addressed include defining physical coverage areas and expected user locations, selecting appropriate frequency allocations and licensing requirements, determining hardware selections for portables, mobiles, repeaters, and infrastructure, layout topological schematics of the repeater and tower locations, design backhaul strategies connecting remote sites, develop console and logging solutions for management, create programming standards for channel plans and protocols, evaluate initial and ongoing costs versus available budget. A well-planned design establishes the technical and policy framework guiding practical deployments.

Permitting and Approvals

Advanced systems usually require various permits and approvals prior to construction. Common examples include filing FCC applications and receiving station licenses for fixed assets, acquiring building and electrical permits for equipment shelters, arranging tower leases and seeking FAA approvals for tall structures, coordinating frequency usage with local coordinating bodies, consulting local municipalities regarding zoning for new sites, obtaining environmental reviews for areas near sensitive zones. Completing the permit process well in advance avoids delays once installation begins. Close agency collaboration helps navigate requirements.

Infrastructure Deployment

The physical network comes next, typically starting with core infrastructure: constructing equipment shelters and mounting repeater systems, installing utility power branches, backup generators if needed, deploying microwave or fiber links between shelters, assembling antennas and transmission lines at tower peaks, testing electrical grounds and lightning protection measures, validating environmental controls maintain equipment temperature. Remote site preparations lay the necessary groundwork before connecting radios. Fiber/microwave testing proves backhaul functionality.

Hardware Installation

Moving to the radio layer, technicians mount mobiles and portables in vehicles per equipment lists, program all devices with correct IDs and encryption settings, enter detailed radio caches like contact lists, test full coverage from all planned usage areas, tune antennas or add risers for coverage holes, train end users on radio protocols prior to field rollout. Thorough configuration and testing at every location catches issues early.

Control Point Setup

dispatch positions require installing console hardware matching system design, configuring logging/notification software integration, connecting to recorded audio archives if applicable, programming console mapping of radio identities, validating interoperability with third-party CAD/RMS, and training dispatchers on radio/console functionality. Proper control point preparation unifies infrastructure control.

Documentation and Training

Comprehensive training materials are developed covering system architecture schematics and topology diagrams, equipment and channel programming guides, standard operating procedures for daily use, maintenance manuals for technicians, encryption management directives if applicable, contact lists, and troubleshooting guidance. Documentation and user certifications complete the rollout process with informed end users and administrators.

Repeater

One of the most important technologies enabling advanced radio communication is the use of repeaters. A repeater expands the range and functionality of radios by receiving radio signals and retransmitting them, effectively doubling or tripling the coverage area of a basic transmission. This allows wider area communication networks to be established.

What is a Repeater?

A repeater is an electronic device that receives a radio signal on one frequency band and automatically retransmits it on a different frequency band. It consists of a receiver operating on the input frequency and a transmitter operating on the output frequency, connected to an automatic switching circuit. When an input signal is sensed, it triggers the transmitter to simultaneously rebroadcast the same message.

Typical Configuration

Repeater stations are usually situated at elevated locations to maximize coverage, such as atop tall towers or buildings. The input and output antennas are placed to provide good reception and transmission coverage to the intended service area without overlap that could cause interference or feedback loops. Additional components include a duplexer to isolate the transmitter/receiver, an encoder/decoder if digital, and power supplies with backup batteries.

Operating Modes

Repeaters work in half-duplex mode, only transmitting when not receiving and vice versa to avoid feedback. Common offset configurations include +600KHz (transmit higher than receive) or -600KHz splits to allow simplex radios to use the repeater without adjustments. Time-outs and coding also prevent locking up from continuous transmissions.

Repeater Benefits

The key benefits repeaters provide include extending communication ranges, improving indoor coverage penetration, allowing mobility across wider areas, and enabling additional data functions beyond basic radios. By receiving weak signals and rebroadcasting at higher power, repeaters facilitate reliable links between users separated by distances, terrain features, or structures that a simplex setup could not traverse reliably.

Network Components

Other important elements that comprise a full repeater system include discreet link radios to interconnect repeater locations, backup power generators, equipment shelters, precision antennas, transmission lines, automatic controllers, and monitoring software. Ensuring all components are properly integrated, maintained, and accessible is critical for dependable network uptime meeting user needs.

Digital Capability

While originally analog, today's repeaters fully support robust digital protocols, too. Employing encoding allows additional features like individual unit IDs, status messages, GPS data, text communications, and even AVL/telemetry integration. Digital operation renders the network less susceptible to interference while expanding versatility beyond basic voice calls alone.

Repeaters in Practice

Some practical implementations of repeaters include linking mountain towns to surrounding areas, providing maritime ports coverage offshore, spanning rural territories for first responders and utilities, augmenting cellular "dark spots," tying together tunnel/mine complexes, and serving as the backbone for large scale P25 and DMR radio networks. Wherever extended communication ranges are required, repeaters deliver robust solutions.

Programming

Proper programming of all involved radios is crucial for successful repeater utilization. Technicians must input the correct receive frequency, transmit frequency offset, tones (CTCSS/DCS), radio IDs, and operating modes according to established standards. Comprehensive documentation assists users and aids in troubleshooting potential issues down the line.

Ongoing Maintenance

Like any critical network component, repeaters demand ongoing administration and maintenance. Common tasks involve monitoring performance logs, conducting preventative equipment checks, replacing end-of-life components, updating software/firmware, servicing backup power, trimming vegetation, inspecting shelters/towers, and more. Proactive upkeep maintains uptime expectations.

Signal Analysis

Engineers also routinely conduct signal surveys and coverage modeling to validate network parameters and identify weak spots requiring attention. Drive testing with portable radios and RF probes helps tune antennas, confirm quiet areas aren't dropped calls, and reoptimize for changing user/terrain conditions over time.

Future Possibilities

Emerging repeater technologies on the horizon include Software Defined Radio for versatile multi-protocol support, wideband designs for combining adjacent channels, advanced beam forming to tighten coverage lobes, Automatic Repeater Addressing for seamless wide-area network registration, onboard digitization assisting with encryption/error correction, and use of unmanned drones as quickly deployable aerial repeaters after disasters. Constant evolution fuels expanded capabilities.

Benefits of Using Advanced Repeater Servers

While basic repeaters effectively extend radio ranges, modern advanced servers take repeater functionality to new levels. Server-based repeaters employ computer control and networking to deliver enhanced capabilities beyond standalone units. This provides significant advantages for supporting robust, feature-rich communication systems.

Remote Management

A core benefit is remote management and monitoring capabilities. Server platforms allow remote access and configuration via secure VPN/IP connections. Administrators can make settings changes, software updates, diagnostics, and more without visiting sites - a major time savings. Integrated control software also monitors all repeaters/sites from any central location. Alerts indicate issues needing attention.

Expanded Control

Advanced servers integrate robust Linux/Windows control software, empowering more than simple on/off controls. Features include integrated scheduling, telemetry, logging, conditional logic, and a fully scriptable environment. This allows custom automation like remote resets, automatic firmware updates based on conditions, detailed repeater behavior programming, and much more. Possibilities are only limited by imagination and expertise.

Network Scalability

While standalone units connect individual locations, server infrastructure facilitates linking entire networks of interlinked sites. This expands coordination potential to county-wide, regional, or even continent-scale frameworks. Larger networks distribute traffic intelligently, interconnect systems across bands/agencies, and escalate to optimal paths, enhancing reliability. Server control maintains cohesiveness as the footprint grows.

Integrated Services

Servers also consolidate valuable services onto a common infrastructure. Examples include integrated VoIP telephone interconnects, remote base/fixed control stations, WiFi access points, surveillance cameras, environmental sensors, and more. Consolidating reduces components while increasing integrated functionality across sites.

Robust Connectivity

Server-enabled networks support robust high-speed backhaul strategies beyond point-to-point radio. Options include high-bandwidth Ethernet, cellular, microwave, and satellite techniques. This bandwidth boosts payload capacities for streaming video, large file transfers, telemetry systems, mesh networking, and future capacity demands. Reliable backbones benefit resource coordination.

Enhanced Security

Computer-controlled infrastructure facilitates robust user/group permissions, authentications, VPN infrastructure, encryption protocols, malware protection, and traffic monitoring expected of networked gear. Tight security standards protect sensitive coordination and strategically minimize attack surfaces compared to exposed standalone devices. Auditing also enables compliance proof.

Redundancy

Servers facilitate the deployment of redundant hardware, power supplies, and networking to ensure uptime reliability on par with mission-critical utility infrastructure. Features like failover routing between servers, automatic backup power activations, and geographically diverse hosting support reliable operations even under stress. Fewer single points of failure benefit around-the-clock dependability.

Future Evolution

Finally, server foundations supporting programmable logic, robust backhauling, and centralized control pave the way for powerful emerging technologies. Examples include software-defined radios for dynamic spectrum access, narrow banding and wideband capabilities, automated machine learning systems, IoT device integrations, and feature expansions only server-driven networks can harness - translating directly to value-added utilization.

What is Digital Communication

Communication systems have progressed tremendously from early analog methods to modern digital techniques. While analog transmission conveys data as continuous signal variations, digital methods instead represent information as discrete binary values.

Analog vs Digital Signals

In analog signaling, the transmission directly corresponds to the actual signal waveform, such as varying voltages representing sound waves. Digital techniques instead sample analog values at specific intervals, quantizing them into numeric representations using a finite set of levels. For example, a voice waveform might be sampled 8000 times per second and coded using 256 different voltage levels.

Digitization Process

Sampling means measuring the analog amplitude at frequent, uniform time intervals. Quantization then assigns the measured level to the nearest value within the available resolution. For audio, 8-bit samples provide 256 levels, while higher resolutions like 16-bit increase fidelity. Digital signals can then reconstruct the original via digital-to-analog conversion for playback.

Binary Encoding

each discrete quantization level is assigned a unique binary code comprised of 1s and 0s. Common audio formats like MP3 leverage psychoacoustics to optimize compression by removing inaudible components from the binary stream. Video digitization handles multiplexed audio/visual channels at frequencies matching human perception. Proper encoding balances quality and bitrate needs.

Channel Coding

To maximize reliable transmission, source encoding alone leaves signals vulnerable to corruption. Channel coding adds structured redundancy to detected binary patterns, enabling error correction. Techniques range from simple repetition coding to sophisticated block/convolutional schemes systematically reincorporating checksum information. Coding trades bandwidth versus noise immunity.

Modulation

Transmitting raw digital signals wastes bandwidth. Modulation instead maps binary data to properties of a higher-frequency analog carrier wave suitable for transmission. Popular schemes include amplitude-shift keying representing 1s and 0s by signal amplitude, frequency-shift keying using discrete frequencies, and phase-shift keying relying on phase differences. Modulation optimizes transmission characteristics.

Multiple Access

When multiple simultaneous transmissions share the same medium, multiplexing techniques organize who can transmit and when to avoid interference. Methods include frequency-division multiplexing, carving the channel into frequency slices, time-division systems taking turns within a timeslot framework, and code-division approaches uniquely encoding non-colliding spread spectrum signals. Proper selection matches usage profiles.

Networking

Digital infrastructure facilitates networking by segmenting communication into data packets and identifying source, destination, and content. Switches and routers then intelligently direct traffic through interconnected nodes based on network protocol headers. Advanced techniques like multipath routing optimize path selection and quality of service features to prioritize latency-critical traffic. Networks scale transmission resources and coordinate devices.

Receiver Operation

At the destination, the receiver follows the inverse process of modulation, channel decoding, demultiplexing, and digital-to-analog conversion, recovering the original signal in a usable form. Synchronization ensures that transmit/receive nodes coordinate sample timings during reception. Equalization counters distortion to restore signal fidelity as close as possible to the original pre-transmission state.

Advantages

Key benefits of digital communication include immunity to noise, interference, and attenuations, which degrade analog fidelity over distances. Digital encoding also facilitates error correction, multiple access, networking, information security, signal processing, and multimedia convergence beyond what's possible through analog-only means. Advanced error correction techniques ensure reliable transmission even under hostile link conditions.

Applications

Wide-ranging applications of digital communication now impact daily life. Examples include mobile/cellular telephony, internet networking, TV/radio broadcasting, satellite services, Wi-Fi connectivity, Bluetooth, GPS, two-way radios, and transportation technologies like toll collection. As bandwidth increases and rates decrease, digital ubiquity will only continue growing across industries and technologies riding its backbone. Advancements in areas like 5G, fiber optics, IoT, and integrated networking portend even more possibilities.

Features of Digital Communication

The transition from analog to digital transmission unlocked entirely new possibilities for robust, capable communication networks. Beyond simply representing signals numerically, digital incorporates supplementary techniques to optimize transmission characteristics and functionality.

Discrete Levels

The fundamental attribute is a representation of information as discrete levels rather than a continuous spectrum. Digital assigns samples to a finite set of numeric values, in contrast to analog's limitless gradations. This quantization underlies binary encoding and coding/decoding processes anchoring digital operation.

Binary Encoding

All digital data and signals are ultimately expressed as sequences of binary digits - the electric or optical pulses representing the numerals 1 and 0. Mappers convert discrete levels into unique bit patterns for transmission in bandwidth-efficient streams. Advanced codes also embed error detection data.

Synchronization

Coordinating transmit/receive timing ensures aligned sampling of binary bitstreams. Techniques like preamble sequences, pilot tones, and frame synchronization patterns cue receivers on boundaries between packets, symbols, coding blocks, etc., for accurate decoding.

Error Detection/Correction

Systematic coding adds controlled redundancy, enabling the detection and correction of bit errors introduced during transmission. Strategies range from simple parity checks to powerful convolutional/Turbo product codes. This resilience counters attenuations degrading analog quality over links.

Modulation

Binary encodings shape baseband signals unsuited for efficient transmission. Modulation encodes bits as properties of carrier waves like amplitude, frequency, or phase shifts well-adapted to specific media. Processes at receivers demodulate signals back to the initial binary level for processing.

Multiple Access

Techniques organizing concurrent transmissions to avoid collisions include frequency-division multiplexing dividing bands, time-division systems taking scheduled timeslots, and spread-spectrum coding spread signals across the spectrum. Standards dictate protocols

Packetization

Data is packetized by bundling bits into frames delineating sources, destinations, content lengths, and more. Standard protocols establish well-defined rules for networking equipment to reliably direct traffic through interconnected nodes to proper endpoints.

Addressing

Unique identifiers like media access control addresses, IP numbers, domain names, and port assignments embedded in packet headers coordinate delivery to specific networked devices or services regardless of location on interconnected infrastructure.

Networking

The combination of packet switching, addressing, and standardized protocols enables robust digital network scaling connectivity via physical/wireless links between interconnected nodes. Central routers direct traffic optimally through topology.

Signal Processing

Digital facilitates signal processing like encoding/decoding, modulation/demodulation, filtering, equalization, and error correction through programmable algorithms operating on discrete numeric representations. Flexibility optimizes performance for diverse applications.

Data Compression

Lossy and lossless statistical/predictive compression algorithms remove spatial/temporal redundancies from signals like images, video, and audio to optimize throughput for bandwidth-limited media. Standards balance quality versus efficiency.

Information Security

Cryptographic ciphers scramble signals using keys during digital transmission for privacy and authentication on networks. Techniques range from basic XOR operations to advanced asymmetrically secure public/private key exchanges.

Quality of Service

Network features prioritize time-sensitive services like VOIP, video calling, and remote surgery over bandwidth loads. Techniques include traffic shaping, committed access rate, class of service, and prioritization to ensure critical functions don't suffer latency.

Multimedia Convergence

The combining voice, video, text, and multimedia under a common digital framework has enabled unified services across a variety of devices and media, from televisions to mobile phones to smart speakers.

The Advancements in Satellite Communication Technology

Since the launch of the first artificial satellite, Sputnik 1, in 1957, satellite communications have revolutionized how we connect across vast distances. Each new generation of satellites brings higher capacities, lower latencies, and more resilient services. This section will examine some of the major technical advancements that have propelled satellite communications into powerful global infrastructures supporting modern societies.

Higher Throughput Satellites

Early communication satellites in geostationary orbit relied on single channel per carrier (SCPC) designs with limited capacity. Later, satellites increased power and adopted Time Division Multiple Access (TDMA), dividing each transponder into timeslots. Newer High Throughput Satellites (HTS) employ spot beam technologies, concentrating bandwidth over specific geographic areas. Massive

MIMO antenna arrays form thousands of independent beams, dramatically multiplying throughput. Modern satellites now deliver terabits per second (Tbsp.) of aggregate capacity to backbones worldwide.

Smaller Satellites

Miniaturization enabled new classes of smaller, lower-cost satellites. Early examples include the USAF's Project West Ford, which expelled clouds of nickel needles in 1961. Later, smallsats employed off-the-shelf electronics, unleashing the CubeSat standard. Weighing under 1.33kg, CubeSats opened space access through low-cost launch rides. Larger smallsats now provide niche capabilities, while CubeSats demo technologies and perform science missions. Proliferation reduced costs, making space economically feasible at scales unimaginable just decades ago.

Non-Geostationary Orbits

While geostationary orbit reigns as the dominant communications perch, other orbits now fill niches. Medium Earth orbits (MEO) constellations like Iridium lowered latency by locating closer to Earth. Low Earth orbit (LEO) mega-constellations will provide universal broadband coverage through networked swarms with SpaceX's Starlink vanguard. Sun-synchronous polar orbits image the entire planet daily. Diverse orbits optimize services from broadband to earth observation.

Advanced Satellite Technologies

Improvements in power systems, onboard processors, and electronics amplified satellite functionality. Flexible software-defined payloads dynamically shape coverage. Electric propulsion maneuvers satellites without consumables. Onboard digital signal processors pack transponders with advanced modulation. Cognitive radios sense spectrum openings. Regenerative payloads recharge via solar beams. Intersatellite links networked constellations into mesh topologies. Joint government/commercial ventures accelerated innovation cycles.

Integrated Ground Infrastructure

Radiofrequency (RF) and optical ground stations evolved dramatically. Electronically-steered phased arrays compressed spot beams. Frequency-agile up/down converters supported versatile satellites. Fibre backhauling connected remote terminals to internet backbones. Software-defined networking (SDN) virtualized resources. Cloud technologies manage assets via web interfaces. Distributed global networks advanced reliability through redundancy. Integration streamlined operations over consortiums of allied ground station operators.

Broadband Internet Services

While telephony comprised early applications, satellites now deliver vital broadband trunking and consumer internet. High-throughput payloads coupled with ground segment enhancements deliver multi-gigabit connectivity. Low-latency LEO and MEO constellations promise competitive speeds for enterprises and consumers worldwide. Integrated hybrid services seamlessly handoff between satellite, cellular, and terrestrial links. Advanced quality-of-service features prioritize mission-critical communications. Ubiquitous coverage closes the digital divide globally.

Robust Applications

Expand satellite applications include communications on the move for aeronautical and maritime mobility, ubiquitous VSAT networks supporting cellular backhaul/offices/ATMs, television/radio distribution, government networks, emergency response, cellular backhaul/traffic offloading, IoT/M2M connectivity, weather monitoring, earth observation, tracking ships/vehicles and more. Versatile payloads coupled with powerful ground stations enable synergistic services globally.

The Future of Satcom

optical inter-satellite links may interconnect constellations via laser cross-links. Next-generation wideband systems utilizing complex modem/coding techniques could enable terabit throughputs from space. Reconfigurable payloads may support diverse spectrum allocations. Hybrid architectures fusing terrestrial wireless, fiber, and multi-orbit satellite networks point toward globally ubiquitous gigabit connectivity on land, air, and sea. Continued smallsat reductions promise democratized space access. These on-going advancements sustain satellite communications' essential role in a hyper-connected world.

CHAPTER 05

Emergency Preparedness

Emergency Communication

Effective communication is one of the most vital elements of emergency preparedness and response. When disaster strikes and regular means of communication fail, having reliable alternate methods can save lives by allowing the coordination of rescue efforts, requests for aid, and dissemination of critical information to the public.

Radio Hardware

A variety of radio hardware options exist for emergency communication needs. Choosing the right equipment depends on the specific scenario and terrain. Mobile radios installed in vehicles provide communication ability while in transit, a necessity for emergency responders. These should have wide frequency coverage, including VHF and UHF public service bands. Mobile VHF/UHF dual-band or tri-band radios offer maximum flexibility. For base station use at emergency operation centers, look for radios with extra receiving capabilities, large speakers, and cloning features to rapidly program additional radios. Antennas suitable for the frequencies needed should also be selected. 1/4 wave and 5/8 wave "Nagoya" antennas paired with the UV-5R provide good performance for local VHF use. For longer-range communication, higher gain Yagi antennas oriented for directional transmission boost effective radiated power. Magnetic mount and trunk lip mount styles allow for temporary installation on vehicles. Permanent mounts with coax runs keep antennas performing at top efficiency.

Power Supply and Backup Power

Reliability also depends on power sources, as emergency needs may continue for days without utility power. For vehicles, ensure radios can run from the 12V power outlet or cigarette lighter. Deep cycle marine batteries provide a backup that can be recharged by a generator or alternate power source when stationary. Solar panels paired with solar charge controllers and deep cycle batteries create an off-grid solution. Panels from 50-100 watts coupled with 100Ah batteries sustain radio operation.

Organizing Communication Assets - Radio Service Organizations

Organizing communication assets streamlines response. Local radio clubs and emergency response groups coordinate volunteers and resources. Main organizations include:

- ARES (Amateur Radio Emergency Service) - Volunteers from the Amateur Radio Relay League provide communication for disasters and public service events. At the local level, ARES has formal agreements with local public agencies.
- RACES (Radio Amateur Civil Emergency Service) - ARES members can be enrolled in state and city-level RACES to receive emergency powers from served agencies.
- MARS (Military Auxiliary Radio System) -Department of Defense volunteer group providing auxiliary emergency communications.
- CERT (Community Emergency Response Teams) - FEMA program training citizens to prepare for, respond to, and assist in recovery from disasters. Communication support roles available.
- SATERN (Salvation Army Team Emergency Radio Network) - Operates Salvation Army communication networks for disasters alongside ARES/RACES.

Joining local emergency response radio teams opens participation opportunities. Annual registration or enrollment may be required. Attending local emergency operations center practice drills maintains knowledge of procedures specific to the area. Participation allows the utilization of served agencies' repeaters and resources during activations.

Operating Procedures

Knowing standard operating procedures streamlines communication during emergencies. Formal message and traffic handling protocols facilitate coordination. Simplex versus repeater use depends on the situation - simplex preserves repeater resources while repeaters extend range. Proper use of identifiers, phonetics, and plain language keeps exchanges clear and avoids confusion or jamming important traffic. Understanding regulations governing authorized frequencies and power levels ensures legal and effective operation. Adhering to good radio etiquette like brevity, accuracy, and listening first maintains orderly coordination. Pre-scripted templates for common messages expedite call handling. Using common command terminology established by agencies served facilitates coordination with other groups. Advance planning ensures familiarity with procedures to reduce stress and boost focus during rapid events.

Emergency Scenario Practice

Preparing for real emergencies involves practicing communication skills through simulated scenarios. Local radio clubs and response teams regularly hold practice drills to test plans and find areas for improvement. Scenarios can involve simulations of natural disasters like hurricanes, earthquakes, or wildfires, involving evacuation coordination and resource requests among participants. Others stage public service events like marathons or recovery efforts to maintain relationships with served agencies. Individual "go kits" containing relevant maps, station log sheets, and other forms maintain readiness for rapid deployment. Ongoing training ensures familiarity with roles, keeps skills up-to-date, and identifies training needs for new volunteers.

Radio Hardware Standard (Mobile of UHF/VHF Base Radio, Mobile

HF Radio or Base station, VHF/UHF Amateur Radio antennas)

Having reliable and appropriate radio equipment is crucial for effective emergency communication. When selecting radio hardware, it's important to choose options suitable for the frequencies, terrain, and usage cases likely to be encountered.

Mobile VHF/UHF Radios

For emergency responders who need communication capabilities while in transit, mobile dual-band VHF/UHF radios are ideal. They allow real-time voice contact on standard public service frequencies. Recommended radios have wide receiving coverage, including important bands like:

- VHF Low Band (30-50MHz): Used by taxi services, dump trucks, and some public utilities.
- VHF High Band (138-174MHz): Primary public safety bands for police, fire and EMS.
- UHF Low Band (420-450MHz): Used by some public services and industrial applications.
- UHF High Band (450-520MHz): Commercial and federal government radio systems.

Dual-band or tri-band mobile radios like the Icom IC-2730A, Kenwood TK-2609, or Yaesu FTM-100DR provide flexible coverage of these bands from a single unit. Features like wideband receivers, large backlit displays, and front panel programming via USB are convenient during stressful emergencies. Permanent mounting in vehicles prevents radios from becoming projectiles during wrecks. Include power cables allowing use from vehicle power or external batteries.

Mobile HF Radios

For long-range communication beyond the typical VHF/UHF groundwave range, high-frequency (HF) mobile radios offer greater connectivity potential. Popular options include the Icom IC-705, Kenwood TS-2000, and Yaesu FT-991. These radios operate on HF amateur bands from 1.8-30MHz, allowing contacts anywhere in the world to use ionospheric propagation modes. Models have wideband receivers and transmitters, bandwidths suitable for both voice and data. Extra filtering minimizes interference. For permanent installation in emergency response vehicles, mobile mount the radios securely and run coaxial cables to quality discount or whip antennas mounted externally. Magnetic loop antennas provide an alternative with low visibility that mounts unobtrusively under vehicles for stealth applications. For portable use at temporary staging areas, compact all-mode transceivers are convenient.

Base Station VHF/UHF and HF Radios

At fixed emergency operations centers, consider VHF/UHF base radios offering extra performance advantages over mobiles. Models like the Icom IC-2730 and Kenwood TS-2000 have ruggedized chassis, higher power outputs of up to 125 watts, and extra-large front panel controls suitable for noisy EOCs. For HF base stations, transceivers like the Icom IC-7610, Kenwood TS-990S, or Yaesu FT-DX-101MP provide comprehensive all-mode coverage from 160m through 6m bands. Quality external antennas further boost the range. Interfaces like built-in automatic antenna tuners simplify antenna changes.

VHF/UHF Antennas

For reliable voice communication within the typical groundwave range of VHF/UHF systems, 1⁄4 wave and 5⁄8 wave "Nagoya" fiberglass antennas have become standards for portability and effectiveness. Mount these antennas as high as possible on vehicles or portable masts when stationary for best results. Higher gain Yagi antennas improve long-distance performance. Directional 8 to 13-element Yagis cover channels or frequency-specific bands. Rotatable designs allow pointing for maximum signal. For portable use, lightweight fiberglass-built Yagis fold for transport and then erect quickly. Base station antennas require gains sufficient to reliably reach distant repeaters or simplex ranges of 100 miles or more. Vertical antennas like the Comet SBB-5 or Butternut HF6V cover multiple HF and VHF/UHF bands from a single tower or mast installation. For directional communication, Yagi arrays of 15 or more elements boost effective radiated power in desired azimuths.

Power Systems (like solar systems and other off-grid methods)

Reliable power is mission-critical for continuous emergency radio communication. When utility power lines fail during disasters, alternate power sources become essential. This section explores solar power solutions and other off-grid methods recommended to sustain radios and associated IT/monitoring equipment for days or weeks until normal services resume.

Solar Power Systems

Off-grid solar is ideal for emergency communication sites due to silent operation, fuel-free reliability, and scalability. Primary components include photovoltaic (PV) solar panels, charge controllers, batteries, and optional inversion for AC loads.

Panels - Renogy and GoPower solar panels in the 100-300 watt range effectively charge batteries for radios and camp loads. Larger stations use panels up to 1KW mounted on roofs or free-standing frames.

Charge Controllers - MidNite Solar Classical or MNTech controllers precisely regulate the voltage to battery banks for optimal performance while protecting batteries from overcharging. Bluetooth-enabled models monitor charge rates.

Batteries - Deep-cycle marine/RV batteries like Lifeline/Rolls hold charges for weeks without sun. Install 8-16 6-volt batteries with combined capacities of 800-1600 amp-hours for reliable backup power.

Inverters - For powering AC loads, Outback or Sensata pure sine inverters convert battery DC to stable modified sine AC power. Dual-use inverters/chargers allow grid fallback power sourcing.

Typical off-grid household solar systems provide intermittent communication station power. Larger 3KW+ solar arrays hardwired to 4,000+ amp-hour battery banks fuel uninterrupted operation for emergency management facilities and shelters. Redundancy ensures uptime - install backup solar/generator or fuel-cell systems.

Fuel-Powered Generators

Portable gas generators sustain power 6-12+ hours on a single 10-20 gallon fuel tank. Honda EU series generators under 2KW run radios/laptops comfortably. Larger EF-series or Yamaha units power full shelters. Factor ongoing fuel costs/availability, noise, and exhaust fumes. Stationary "whole house" Generac or Kohler diesel/NG generators provide weeks of runtime on hundreds of gallons of stored fuel reserves. Automatic transfer switches switch critical loads to generator power when the utility fails. Consider emission regulations.

Hydro, Wind and Fuel Cells

Small-scale hydro turbines or wind turbines synced to charge controllers/batteries harness renewable energy. Hydro Kensy's small hydro units convert streams into kilowatts. Bergey or WindTronics turbines are installed on masts for consistent power regardless of sun/wind. Consider site suitability. Fuel cells like Goal Zero Yeti generate power from Propane or butane canisters, bypassing the need for solar/wind resources. Goal Zero units provide several days of runtime and recharge via supplemental solar. Plan for fuel resupply. Hybrid "micro-grid" systems optimally combine solar, hydro, fuel cells, and battery/generator backups for redundancy and prolonged off-grid sustainability. Monitor and maintain all system components routinely to ensure functionality during emergencies.

Radio Connectors

Reliable connections between radios, antennas, power sources, and peripherals are just as important as the components themselves for emergency communication networks to function properly. The right connectors ensure components remain attached securely during transport and deployment while withstanding environmental exposures like moisture, heat, vibration, and impacts.

Permanent Installation Connectors

For base stations or vehicles with fixed radio equipment, ruggedized weatherproof connectors maintain solid connections.

Coax-N-type, TNC, and SMA connectors form durable antenna-to-radio connections. N-types handle higher power with machined contacts, while SMA/TNC types work for low-power portable uses or indoor runs. Use quality plated connectors with sealed gaskets.

Power - Anderson PowerPole connectors provide secure, keyed power connections that don't pull apart accidentally. PP15/30/45 sizes accommodate various gauge wires from batteries, solar panels, or generators. Include fuses within 18 inches of battery positives.

Networking - CAT5e and fiber optic runs inside buildings or between comm vans use weatherized fiber connectors or shielded RJ45 jacks securely anchored.

Proper preparation and crimping/soldering ensures solid mechanical and electrical bonds withstanding shocks. Leave connectors accessible for maintenance.

Mobile Radio Connectors

For regular radio/antenna changes in mobile setups, consider fast-attach connectors.

Coax - PL-259 connectors attach quick-mount 5/8 wave "Nagoya" antennas securely. BNC and SMA handle allow portables. Seal all connections thoroughly against moisture.

Power - Use Deutsch "Bullet" connectors, providing positive locking power between radios and vehicle batteries. Include fusible links.

Accessory ports - Use sturdy right-angle USB cables secured against unplugging on radios. Consider waterproof signal ports for data/audio devices.

Portable Radio Connectors

For operations away from vehicles, portable radios require durable, compact connectors:

Antenna - SMA, BNC, or SMB female sockets pair with screw-on portable antennas. Include dust caps when antenna-less for protection.

Audio - 3.5mm TRS connectors attach inexpensive earbuds or shoulder-mic headset cables. Waterproof versions are secure against the weather.

DC power - Rigrunner or Anderson Power micro-connector cables allow portable power banks or 12V adapters.

Data - Micro-USB cables or lightning connectors updater/program radios in the field. Include protectors/caps.

Best Practices & Emergency Procedures

Radio Discipline

Strict adherence to proper radio discipline is essential for effective and orderly emergency communication. Operators must confirm authorization before transmitting on public service channels or repeater inputs. Clear identification using CERT/ARES/callsign and location is required when identifying. The phonetic alphabet and numbers should be used along with plain language for accuracy. Transmissions must remain brief and pertinent to avoid congesting the airwaves unnecessarily. Listen before transmitting to avoid disrupting traffic already in progress. Traffic should be handled politely and respectfully by all operators. Complete monitoring of assigned frequencies prevents missing important information.

Frequencies and Licensing

Compliance with applicable regulations is important for legal and coordinated operations. An FCC amateur radio technician license is needed to access amateur radio bands. Only authorized agencies can utilize restricted public service channels. Transmissions must fall within the permitted output power and frequency ranges. Local repeater frequencies require permission to access as well.

Emergency Communications Plans

Advance planning streamlines response. Emergency plans and standard operating procedures for the local area and served agencies must be thoroughly understood. Assignments, roles, and the communication organization structure must be clear. Response checklists provide needed reminders. Advance familiarity with required forms, message formats, and record-keeping protocols prevents delays. Establishing operations procedures in cooperation with partner organizations facilitates coordination.

Network Operations

Effectively managing response requires properly deploying available resources. Mobile and portable radios should be outfitted for temporary staging areas and deployed teams as the scale of the incident dictates. Rotating radio use or operating shift schedules helps maintain continuous coverage by allowing equipment to charge between periods of transmission. All traffic, messages, and reports transmitted must be recorded to maintain an accurate operational record. Sheltering equipment protects it from hazardous weather conditions, while securing it safeguards radios from theft or tampering. Disseminating pertinent non-traffic information across the wider communication network aids situational awareness.

Proper Radio Equipment Usage

Optimal performance relies on correctly utilizing equipment according to design standards. Antennas are most efficiently deployed using recommended mounts, cabling methods, and orientation. Programming radios with accurate frequencies, tones, and labels ahead of time avoids wasted set-up time during an incident. Bringing backup radios, portable power sources, and charging gear provides redundancy in case of component failure in the field. Operators must strictly follow manufacturer instructions and adhere to guidelines on equipment limitations, safety precautions, handling, and maintenance. Upon clearing from the operational area, all communication assets must be restored to full working order and properly charged or stored.

Scenario Training

Realistic scenario-based training proves essential. Participation in annual mock emergency exercises tests all procedures under realistic activations. Subjecting equipment to full operational tests and documenting any identified issues allows prompt correction. New volunteer training imparted through mentoring establishes clear expectations and ensures protocols 传达 clearly. Conducting after-action reviews incorporates lessons identified during drills to refine plans and closing preparedness gaps. Continuous improvement through training enhances response performance in actual incidents.

Seven Baofeng UV-5R Survival Tips Off Grid

The affordable and feature-packed Baofeng UV-5R handheld radio has become a popular choice for emergency preparedness. When utility infrastructure fails and normal means of connection disappear, the UV-5R offers indispensable voice communication ability.

Tip 1: Compile Frequency Lists

Programming radio memories ahead of time streamlines operations under pressure. Save critical frequencies on local repeaters and simplex channels, as well as national interoperability frequencies. Databases like RadioReference. com provide crucial coordination details. Have multiple programmed backup radios ready.

Tip 2: Assemble Accessories

Pair the UV-5R with recommended attachments - Nagoya NA-771 antennas boost range. Extras include headsets, portable power banks, and protective cases. Include charging cables, a manual, and a waterproof notepad/pen for logging contacts. Spare batteries let radios run non-stop. All gear fits within a protective portable kit.

Tip 3: Practice Portable Operations

Emergencies find many away from vehicles and infrastructure. Deploy the UV-5R effectively by setting up a temporary antenna and power source. Drill up deploying antennas like the Nagoya efficiently. Have tricks for rigging improvised antennas. Know battery-saving methods like sleep modes to maximize runtime away from charging.

Tip 4: Optimize Settings

Default radio settings don't always maximize performance. Tweaks like raising power levels safely, enabling vox features with headsets, and adjusting squelch remove unnecessary noise. Use coding squelch to access repeaters requiring PL tones privately. Ensure display clarity in sunlight with adequate contrast and backlight levels.

Tip 5: Establish Communication Plans

Determine evacuation routes and pre-set regular check-in frequencies and times with contacts. Have a call plan for emergencies requiring relaying critical information. Establish a call sign convention, making assisted parties easily identifiable to increase coordination speed. Memorize important contact details ahead of time.

Tip 6: Shelter Equipment Strategically

Prevent radios from total failure due to heat, cold, moisture, or impact. Boil-proof zip bags encapsulate components. Hard plastic cases shield delicate controls. Programmed spares stay in multiple climate-controlled locations to ensure access regardless of conditions. Mount mobiles securely away from debris paths.

Tip 7: Maintain Proficiency Regularly

Without routine use, radio skill decay occurs precisely when maximum competency becomes critical. Schedule regular test communications and scenario training to retain operational excellence. Keep firmware and documentation updated online and offline. Review operating standards and consider augmenting them with formal instruction. Routine assesses radio condition early before total failure occurs.

Amateur Radio Emergency Organizations

When disasters strike, coordinated volunteer communication networks amplify emergency response capabilities.

ARES - Amateur Radio Emergency Service

As the official emergency arm of the ARRL, ARES recruits radio amateurs for public service events and disaster response. Local ARES groups develop Standard Operating Procedures aligning with served agencies like the Red Cross, local emergency managers, and national ARES plans. Members train for roles and assist when activated by agencies. Permissions allow operation on restricted frequencies. The structure includes local Emergency Coordinators, Assistant ECs, and public service managers coordinating volunteers. Members register NIMS credentials with FEMA, providing consistent communications when other systems fail. ARES emphasizes operational readiness through ongoing training, drills, and maintaining portable equipment caches. Annual Field Day exercises involve emergency power generation and deployment practice.

RACES - Radio Amateur Civil Emergency Service

Operated state-to-state, RACES originated as the civil defense arm of amateur radio during WWII. Members register identification numbers, allowing operations on authorized government frequencies during actual emergencies or exercises. Local RACES organizations exist under civil defense/emergency management offices. They activate to provide health and welfare traffic handling or other required communication when public systems become overwhelmed. Many RACES members also participate in ARES to maintain licensing and practice skills more frequently. They require less training than paid staffers but provide auxiliary communication depth, allowing priority systems to focus on emergency response tasks. Official identification permits access to restricted resources as needs arise.

SATERN - Salvation Army Team Emergency Radio Network

Originating in 1953, SATERN members assist The Salvation Army with emergency and disaster communication services. They deploy portable and mobile stations for health and welfare inquiries, equipment transportation coordination, and other aid during disasters. SATERN trains regularly with The Salvation Army to understand their operating structures and support roles. Local teams coordinate food, clothing, and emotional/spiritual care resources for those in need alongside trained staff members. HF voice and digital modes reach distant corps to request or transport personnel and supplies speedily. Emcomm skills allow adapting to shelter, feeding, and distribution point communication demands. Membership requires passing a licensing exam on The Salvation Army's structure. They provide critical connectivity backing when need exceeds normal agency capacities.

SKYWARN - National Weather Service Storm Spotter Program

SKYWARN recruits, trains, and certifies severe weather spotters to identify and describe developing storms. Reports aid meteorologists with real-time tracking, helping save lives through early warning. Spotters deploy to vulnerable areas during threatening conditions armed with guidelines for objectively reporting cloud descriptions, storm impacts, and hail/wind estimations. Digital and voice modes rapidly convey ground-truth to forecast offices. Annual spring training classes introduce spotters to storm structure and behavior, taught by NWS specialists. Classes educate on identifying storm attributes indicating tornadic potential. Spotters proving adept gain enhanced grid square coverage permissions for contributing experienced real-time data. Peer training and Net Control practices spread proficiency to serve frontline NWS needs

CHAPTER 6

Military Communication

Communication Strategy

Effective communication is vital for mission success in any military operation. Developing a comprehensive strategy establishes clear procedures to coordinate activities between units and prevent misunderstandings.

Objectives and Importance of Communication

The first step is defining objectives for the mission or period of operations. Military communication serves purposes including command and control, coordinating efforts, status updates, support requests, and intelligence sharing. Outlining objectives provides structure and ensures the strategy sufficiently addresses communication's key roles. Leadership must clearly explain the importance of effective information flow to subordinates.

Designation of Responsibilities

Following objectives, the strategy designates roles for communication aspects. This includes identifying commanders and staff overseeing networks, appointing signal officers to manage transmissions, distributing devices and materials based on roles and locations, assigning units to handle courier duties, establishing procedures for routine reports and requests, and designating alternates in critical roles. Roles and responsibilities must be clearly defined and disseminated.

Communication Channels

Once leadership and operational duties are identified, the strategy allocates channels for message types and command levels. This includes designating radio frequencies, encryption, and call signs for contact between headquarters and units. Separate radio networks coordinate between companies. Back channels covertly share intelligence. Wire networks, field phones, and runners connect headquarters. Cellular and satellite phones serve as backups. Visual signals like smoke and flags enable emergency contact if electronics fail. Channels must be clear to avoid crossed signals.

Coordination Procedures

Additionally, procedures coordinate ongoing operations routinely and in emergencies. Check-in schedules via radio maintain contact between dispersed teams. Unit codenames facilitate anonymous communication behind enemy lines. Standardized message formatting prevents errors. Read-back procedures verify critical message receipt and comprehension. Emergency codes request reinforcements, medical evacuation, or air support. Contingency plans account for primary channel disruption with tested backups. Procedures ensure coordination across unexpected events and hierarchy levels.

Encrypted Communication

All channels incorporate encryption and security protocols protecting sensitive data from interception. Encrypted radios, phones, and modems issue encryption keys. Rotating keys through bulletins prevents code-breaking over time. One-time pads, codebooks, and algorithms maximize confusion. Anonymous call signs avoid providing intelligence when intercepted alone. Codebook transport and destruction protocols protect compromised schedules. Civilian network vetting avoids signal leaks. Jamming defeats monitoring during sensitive operations. Comprehensive encryption safeguards information advantage.

Command and Control

The strategy delineates command utilization of networks for effective control. Clear succession grounds bypass incapacitated leaders. Digital and analog capabilities rapidly disseminate time-critical orders. Message read-back confirms comprehension, preventing errors. Logging transmitted orders in classified records facilitates documentation, tracking, and cross-checking. Priority marking expedites response. Subordinate briefings on intent allow decentralized decision-making when isolated. Continuous monitoring maintains flexibility in responding to developments requiring redirection. Reliable communication enables strategic leadership.

Evaluation and Improvement

Regular evaluation and adjustment keep the evolving strategy matching realities. Assessing communication performance after exercises and real operations identifies strengths and weaknesses for refining roles, protocols, backups, and encryption practices. Technological advances may necessitate channel or encryption updates. Changing missions influence objective priorities. Feedback from all levels improves future strategies. An iterative approach maintains maximum coordination ability to accomplish varied challenges.

Plan For Communication (Why, When, and How To)

Effective communication is critical for any military operation to be successful. However, communication cannot simply be left to chance - it requires careful planning. This chapter will explore why developing a plan for communication is so important when, in the operational process, planning should take place and how to construct an effective communication plan. With a well-thought-out plan, commanders can ensure their forces have the right information at the right time to accomplish the mission.

Why Plan Communication?

There are several compelling reasons why taking the time to plan out communication is essential for military operations. First, communication enables command and control. Senior leaders must be able to provide strategic direction, issue orders, and adjust the plan as needed based on the evolving situation. Second, coordination between different units and supporting arms requires clear communication. Whether it's synchronized movements, artillery and air support, medical evacuation, or resupply, units must seamlessly pass information back and forth to work as a cohesive, combined arms team. Third, situational awareness relies upon routine status reports from units in the field. Leadership needs updated knowledge of unit positions, casualties, supplies, and other operational details to maintain an accurate, common operating picture. Fourth, unexpected developments often arise that demand rapid communication of new instructions. Fifth, operating across various levels of the chain of command introduces complexity if procedures are not standardized. In every case, a methodical plan laying the groundwork in advance serves as the foundation for effective information exchange throughout the operation. Communication does not succeed by chance alone.

When to Plan Communication

The optimal time to design a comprehensive communication framework is during the planning process for an operation or campaign, well before execution. It is most strategic to consider connectivity concurrently with other mission-related details. During initial concept development, leaders identify overall objectives and concepts of operation. This stage sets the tone for information flows required to enable various components. As mission analysis and course of action development unfold, specific tasks and responsibilities emerge, along with the projected operational environment and potential frictions. These factors illuminate communication mediums and capacities demanded.

Staff planning assigns units to roles, designates assembly areas and objectives, and maps out logistics - all impacting coordination issues to address with connectivity solutions. Orders development, finalizes execution timelines, and specifies subordinate missions. Communication schedules and connectivity resources needed to accomplish sequential steps come into focus. Pre-operation rehearsals allow revising, testing, and confirming communication tactics, techniques, and procedures well in advance of D-Day. Surprises and unexpected demands are minimized.

How to Plan Communication

With an awareness of why and when plan, attention turns to how to develop a comprehensive, detailed communication scheme. Several key factors must be considered:

- Objectives: Define precise purposes like command/control, coordination, reports, and contingency needs.
- Organization: Assign specific nodes, specialties, equipment, and infrastructure roles to clearly delineated communication staff sections.

- Technology: Designate primary and backup platforms considering transmission types/frequencies, encryption, transmission security, and interoperability.
- Protocols: Standard operating procedures and detailed message formatting establish coordination, discipline, and replication.
- Contingencies: Develop alternative connectivity solutions and readiness testing for degraded/denied scenarios.
- Multinational: Anticipate procedures to synchronize with allied communication systems.
- Change management: Account for adjustments necessitated by staff rotations, equipment failures, or modified operations.
- Training: Schedule leader familiarization and competency rehearsals to validate proficiency.

19 Important Principles of Effective Communication

Effective communication is essential for mission success in any military operation. While technology and mediums may change, core principles of communicating remain constant. We will discuss 19 key principles that should guide all military communications to ensure important information is conveyed accurately and efficiently. Adherence to these principles facilitates understanding and coordination.

Clarity

One of the most fundamental principles is communicating with clarity. Messages must be easily understandable the first time and leave no room for incorrect interpretation. Jargon and ambiguity should be avoided, especially in high-pressure situations. The intent or directive should be apparent upon first receipt of the transmission. This reduces delays from having to request clarification.

Brevity

Equally important is brevity. Lengthy, drawn-out communications can lose the attention of listeners and risk confusion. Only information directly relevant to the message should be included. Verbosity does not serve time-sensitive operational communications where clarity and speed matter most. Keep it short while still conveying all critical elements.

Completeness

While keeping it brief, messages must still be complete. All necessary facts, details, requirements, and expected actions should be provided so the receiving party can fully comprehend and respond to instructions without needing additional follow-up. Key information should not be left out due to a desire for brevity alone.

Accuracy

Accuracy is paramount. Factual information, numbers, equipment specifications, and other critical operational details must be double-checked for correctness before transmitting. Errors can seriously undermine coordination efforts or even endanger lives if incorrect situation reports or orders are disseminated. Verify all information for accuracy.

Objectivity

Communication in high-pressure situations requires an objective, unemotional tone. Subjective or emotionally charged language risks misinterpretation and can damage credibility. Stick to impartial reporting of verifiable facts and reasoned requests without injecting personal perspectives or opinions.

Conciseness

Related to brevity, messages aim for maximum clarity with a minimum of words. Eliminate redundant phrases, unnecessary descriptions, and salutations when seconds count. Get directly to the core details needed in as few words as possible while preserving comprehension. Conciseness facilitates rapid consumption.

Composition

Proper formatting and structure help ensure messages are well-organized and coherent for recipients. Start communications with standardized headers indicating the subject or purpose. Group similar pieces of information together logically. Follow consistent protocols for construction wherever possible to streamline understanding.

Standardization

Adhering to standardized terminology, abbreviations, radio codes, and acronyms agreed upon beforehand prevents miscommunication deriving from diverse interpretations of non-uniform language. The common ground keeps all parties aligned without confusion from variable individual usage of technical terms.

Prioritization

Designating transmission priority permits receivers to swiftly interpret urgency and process messages accordingly. Distinguish critical time-sensitive traffic from routine reports. Indicate life-safety messages or orders necessitating immediate action versus general updates. Prioritization helps prevent critical data from getting mishandled or delayed.

Redundancy

Communicating important information through multiple redundant pathways hedges against the potential for single points of failure disrupting connectivity. Parallel voice and digital transmissions, use of alternate frequencies or hard copy backups for encryption keys reduce vulnerability and increase reliability of delivery for mission-essential updates.

Confirmation

Requesting confirmation of message receipt and comprehension, especially for critical instructions, mitigates the risks of data being missed, received incorrectly, or requiring clarification down the line. Read-backs and acknowledgments validate understanding on both ends and trigger follow-ups if needed. Double-checking prevents errors from propagating.

Coordination

Clear designation of coordination requirements—"coordinate with Company A," "send resupply trucks," etc.—furnish contextual cues for recipients' follow-on actions. Specifying coordination delivers needed for streamlined collaboration rather than ambiguous data drops risking initiative stagnation or cross talk. Coordination fosters cooperation.

Relevance

Information exchanged should directly pertain to the stated purpose or operational tasks at hand. Messages bombarding limited communications bandwidth with tangential data distract from mission-essential updates and potentially delay critical information. Keep all traffic pertinent to current objectives.

Security

Classifying data appropriately plus utilizing encryption protects sensitive intelligence and tactical details from enemy exploitation while still enabling friendly coordination. Over-classification causing unnecessarily limited sharing and encryption hindering collaboration should also be avoided, but security measures shield key advantages.

Simplicity

Complex, convoluted messages risk dilution or misunderstanding of core issues under time pressure. Distill information down to its basic, straightforward elements, stating the most germane facts without unnecessary background or elaboration. Simplicity supports swift comprehension.

Pushing Information

Proactively communicating new developments expedites shared situational awareness. Leaders "pushing" relevant updates to subordinates prevents stagnation compared to a pure "pull" model where all data must be explicitly requested. Push alerts subordinate initiative for rapid decisions when flexibility matters.

Audience Consideration

Tailoring message contents and technical terms to the expected recipients' capabilities and needs fosters understanding. Speaking to infantry experience versus headquarters-level factors requires adjustment. Consider diverse ranks, functions, languages, or nationalities when composing traffic for collective situational awareness.

Conciseness

Messages should aim to convey information in the most efficient manner possible using the fewest words. Rambling messages with unnecessary details or repetition slow understanding and take up valuable bandwidth. Getting straight to the point in a condensed yet still clear form allows for better comprehension under time pressures.

Feedback

Seeking confirmation that a message was correctly received and understood is imperative, especially for critical instructions or time-sensitive reports. Asking the recipient to repeat or acknowledge the key details verifies comprehension on both ends and prompts follow-up if needed to clear up any misunderstandings before they impact coordination. Feedback closes the communication loop.

Communication Encryption and Protection

Effective communication is critical for mission success but leaves military forces vulnerable if intercepted by adversaries. Encryption and security protocols shield valuable intelligence and prevent exploited weaknesses. We will discuss why encryption matters, common encryption methods, transmission security, and other protective considerations for safeguarding vital information exchanges.

Importance of Encryption

The most basic reason encryption is indispensable concerns operational security (OPSEC). Revealed details about friendly unit positions, capabilities, staffing, or intentions jeopardize tactical advantage and endanger lives if intercepted by hostiles. Legal compliance also motivates encryption to avoid transmitting identifying information that might infringe on civil liberties if exposed. Signal intelligence collected for national defense purposes requires careful handling. Additionally, encrypted backups hedge against compromised primary communication means. Even minor technical failures or momentary disruptions endanger missions without redundantly encrypted continuity. Finally, alliances require interoperable encryption synchronizing multinational coordination efforts while blocking adversary comprehension. Operational collaboration depends on secure, unified systems. Overall, encryption shields assets central to success and stewardship.

Encryption Methods

A diverse toolkit suits varied connectivity environments. Manual cryptographic systems like one-time pads generate unbreakable random characters mapping to plaintext. Modern alternatives include FIPS-validated commercial solutions applying symmetrical or asymmetrical algorithms upon authentication. Steganography hides messages within innocuous-looking digital files like images or audio, thwarting detection. Frequency-hopping and direct-sequence spread spectrum transmission jumping rapidly between frequencies complicates eavesdropping. Quantum encryption exploits subatomic phenomena promising uncrackable substitution ciphers, while blockchain-distributed ledgers authenticate parties through decentralized consensus. Diverse solutions equip flexible security for any situation.

Transmission Security

Physical transmission security (TRANSEC) complements encryption. Strict emission control and reduced use conserve limited bandwidth, avoiding pattern exposure. Dummy traffic obscures real message timings. Low probability of intercept antennas and directional transmissions narrow broadcast footprints. Mobility evades fixed direction finding. Short, discretely timed bursts replace longer vulnerable signals. Frequency agility and synchronization code jumping confound interception attempts. Jamming interrupts hostile receivers. A layered, multifaceted approach forms a comprehensive protective shield.

Key Management

Secure key generation, distribution, and storage handle encryption "keys," unlocking protected transmissions. Remote rekeying updates cipher machines without handling physical media. Steganographic carrier waves clandestinely ferry replaced keys. Biometrics or machine learning authenticates users. Redundant multi-factor authentication prevents single exploits from compromising all systems. Audits monitor entitlements and revoke access as roles change. Diverse redundant measures prevent any single point of failure.

Planning Considerations

Protection must be integrated into the planning process. Operational needs, regulations, and cross-domain interoperability with partners inform tailored technical solutions. Infrastructure, staffing, and budget reflect necessary capabilities. Contingency procedures rehearse degraded connectivity scenarios. Training standardizes proficiency. Periodic review evaluates technology upgrades, changes to protected information classifications, and revised threats. Dynamic security evolution supports evolving operational realities.

Information Assurance

Beyond encryption, comprehensive information assurance safeguards data throughout its lifecycle with physical security, access controls, and cyber hygiene. Storage follows classification specifications. Secure disposal destroys unusable remnants. Cyber insider threats complement external risks, necessitating multilayered defenses.

CHAPTER 07

Real-Life Scenarios & Case Studies

General Baofeng Radio Use in Real-Life Survival Situations

In any survival scenario, communication is key. Whether facing an emergency, natural disaster, or other dangerous situations, having the ability to call for help or coordinate with others can mean the difference between life and death. The Baofeng radio is an affordable, versatile, and practical tool that has proven invaluable for real people in real survival situations around the world. It's the thread that can guide rescuers to your location, keep you informed about impending danger, or simply connect you with others to coordinate efforts and share critical information. It's often the case that when facing an emergency, natural disaster, or other perilous situations, having the ability to call for help or coordinate with a group can be the deciding factor between a safe outcome and tragedy. The Baofeng radio has emerged as an affordable, versatile, and practical tool that has proven invaluable for real people in real survival situations around the world. The utility of the Baofeng radio in crisis cannot be overstated. In the chaos that follows a disaster, traditional lines of communication are often the first to fail. Cell phone networks become overloaded, landlines are disrupted, and internet access becomes a distant memory. Here is where the Baofeng radio shines, offering a robust alternative that relies on radio frequencies, which are often the last standing means of communication in a dire situation.

This radio is not just a piece of equipment; it's a beacon of hope for those who find themselves in peril. Stories abound of stranded hikers who have used their radios to call for rescue, of communities that have coordinated relief efforts in the wake of hurricanes using the reliable Baofeng, and of families who have stayed in touch after earthquakes severed all other means of communication. The versatility of the Baofeng radio allows for its use in a wide range of frequencies and settings, making it a must-have in any emergency preparedness kit. Moreover, its practicality extends beyond just the ability to communicate. The Baofeng radio is compact and lightweight, making it easy to carry in a backpack or emergency bag. It is designed to be user-friendly, ensuring that even those with minimal radio experience can operate it under stress. The radio's durability is also a significant factor; it can withstand harsh conditions and the rough handling that often comes with survival situations.

Affordability is another critical aspect of the Baofeng radio that cannot be ignored. For the price of a few meals, one can purchase a device that might very well save lives. This makes it accessible to a broader range of people, ensuring that more individuals can be prepared when disaster strikes. The low cost also allows for the purchase of multiple units, enabling families, groups, or emergency response teams to set up a network of communicators who can work in tandem during a crisis. The radio's versatility is further amplified by its wide range of features. It can be programmed with multiple channels, allowing users to switch between frequencies as needed. The dual-band operation enables monitoring of two different channels simultaneously, which is particularly useful in emergency situations where information may be coming through various sources. With a wide frequency range, the Baofeng radio can access the necessary channels for emergency communication, such as those used by search and rescue teams and amateur radio operators.

In the field, the Baofeng radio's capabilities come to life. It can serve as a means to receive weather alerts, connect with local emergency services, or simply stay informed about the situation at hand. For those with the knowledge to harness its full potential, the radio can even be used to transmit encrypted messages, ensuring privacy and security in sensitive scenarios. The radio is not without its complexities, however. Programming it requires some know-how, and optimizing its settings for specific situations can be a learning process. However, this complexity is also a testament to the radio's adaptability and power. With the Baofeng, one is not just buying a product; they are investing in a versatile tool that can be tailored to a wide array of survival scenarios.

The real-life applications of the Baofeng radio are as diverse as the users who carry it. From the mountaineer who needs to check in with base camp over vast distances to the volunteer disaster responder coordinating supply drops, the radio proves its worth time and again. Its role in guerrilla tactics cannot be understated either; in regions where conflict disrupts infrastructure, the Baofeng radio has become a critical tool for communication and coordination.

To truly understand the power of the Baofeng radio in survival situations, one must look at the stories of those it has helped. There are tales of lost individuals in remote areas who have been located thanks to the signal of their Baofeng radio. There are accounts of amateur radio operators who have provided critical communication pathways during wildfires. There are reports from disaster-stricken cities where the Baofeng radios facilitated coordination when it was needed most.

These stories are not just anecdotes; they are affirmations of the Baofeng radio's role as an essential element of modern survival. As our world becomes increasingly uncertain, with natural disasters seemingly more frequent and severe, the Baofeng radio stands out as a beacon of self-reliance and resilience. It is a tool that empowers individuals to take charge of their safety and to lend a helping hand to others when the traditional lines of communication falter.

In essence, the Baofeng radio represents more than just a communication device; it is a symbol of preparedness, a testament to human ingenuity, and a key player in the narrative of survival in the 21st century. For those looking to fortify their emergency readiness, understand the nature of radio communication, or simply expand their toolkit for when the unexpected occurs, the Baofeng radio is an investment that is worth every cent and more. It is a tool that embodies the spirit of survival: adaptable, resilient, and ultimately, indispensable.

Case Study 1: Lost Hunters Rescued in the Sierra Nevada

Vast tracts of protected wilderness span California's rugged Sierra Nevada mountain range. Each fall, thousands flock to the scenic heights pursuing deer and bear seasons amid brilliant foliage. However, these remote backcountry zones can quickly turn hazardous for any explorers straying from well-traveled circuits. This real account demonstrates how crucial radio contact proved in rescuing two avid hunters who lost their bearings high in the Sierras one crisp October weekend.

The Expedition Into Protected Timberlands

Lifelong friends Eric and Dan eagerly packed for their annual hunting trip, exploring new grounds within the Ansel Adams Wilderness. Well-stocked in gear and provisions, they piled into Eric's truck Friday evening, driving hours up the western slopes. At first light, they shouldered packs deep into seldom-trod forests, hunting signs of trophies. By midday, a thick fog rolled in, muting colors and masking ridgelines. Unsure which gully their camp lay down, confusion set in. After fruitless circling as ebony clouds engulfed the heights, nightfall found them drenched and shivering with no shelter or warmth in sight. Their situation grew dire as temperatures plunged. Thankfully, communication remained their sole advantage.

Attempting Emergency Contact

Eric withdrew a 5-watt UV-5R Baofeng always brought "just in case," knowing cell service ceased well beyond trailhead parking. He began methodically scanning emergency channels normally reserved for official use, praying another party patrolled peripherally. Static pervaded until Channel 5 crackled faintly—a lifeline! After hours of bouncing signals through dripping bows, Eric at last made contact with Madera County SAR volunteers running scheduled radio checks from the valley. Between stammered explanations, Eric conveyed their last discernible coordinates, yet visibility measured mere feet. Rescue seemed near impossible. Volunteers assured keeping signal as long as batteries held, boosting hope amid the mounting crisis. Eric and Dan huddled as near the radio as cold allowed, tuning each whisper for updates and directing potential saviors to their invisible position entombed within the shifting mists.

Mobilizing an Intrepid Rescue Mission

Down in Foothills, volunteers rapidly assembled a response team. With fog shrouding thousands of rugged acres, operators worked through the night, triangulating sporadic bursts from Eric's radio bouncing erratically through the vertical maze. At first light, a CHP helicopter joined equipped with directional antennas homing their transmissions. Pilots struggled to navigate zero-visibility airspace, yet persistence paid off. Just after noon, the crew spotted minute figures stumbling blindly below, shepherding down a relief crew. Medics found Eric and Dan disoriented and hypothermic yet recovering with warm IVs and shelter. Had conditions deteriorated further, the scenario might have ended tragically. Instead, their Baofeng summoned rescuers, delivering them safely from the grip of the engulfing high country gloom.

Preserving Connection in Dire Emergencies

This account underscores radios' critical community benefits, especially in remote locales where cell towers cannot reach. Even with equipment not commercially purchased or sanctioned, citizens cooperatively harnessed radio discipline, maintaining contact until specialized teams arrived. Eric and Dan owed their rescue directly to persistent communication through common citizens and professionals alike, dedicating resources to ensuring every transmission received a response. Sometimes, in backcountry perils, the difference between survival and tragedy rides solely on maintaining lifelines however possible. Here, a small radio proved mighty indeed.

Lessons in Self-Reliance

All adventurers traversing protected lands would be prudent in packing radios plus knowledge of operating them under stress. Modern gadgets fail; topographic interference restricts signals. Radios retain functionality requiring solely batteries while imparting lifesaving self-reliance. Citizens assisting each other through joint monitoring uphold principles of community care critical when official presence lags. While hoping dangers never arise, preparedness strategies like radios empower independence, bolstering chances of safety through cooperation until specialized teams arrive. Their message remains - communication holds power even in civilization's blind zones.

Case Study 2: Coordinating Evacuations During Hurricane Dorian

In early September 2019, residents along Florida's Treasure Coast braced for impact as Category 5 Hurricane Dorian menaced shorelines with unprecedented winds and flooding. When the tempest struck, it exposed vulnerabilities coordinating aid across the fragmented barrier islands comprising Martin County. This account highlights how citizen radio operators helped fill response gaps, assisting evacuations amid infrastructure breakdowns and isolating vulnerable communities.

Building Preparedness Networks

Prior storms taught Leslie, an EMT living on Hutchinson Island, limitations hindering emergency services from reaching everyone alone. She recruited "C.O.M.M. radio" volunteers assisting coordination separately from public systems prone to crashing during disasters. Participants attended training learning medical triage plus radio procedures for reporting needs discreetly without attracting looters.

Monthly drills practicing evacuation roles helped volunteers familiarize themselves with island evacuee populations. County officials supported efforts without endorsing specific equipment for liability. Still, RADIOS improved self-reliance. As Dorian intensified, the network activated, sharing real-time intelligence and liaising directly between residents and responders.

Battering Winds Isolate Communities

On September 1st, Dorian's eyewall slammed the coast as a high-end 4. Pre-dawn winds above 200 mph flattened structures while storm surges isolated whole neighborhoods under raging seas. Communications towers disintegrated, and cell/911 towers fell silent, islanded by raging currents, rendering response impossible via roads now liquid highways. C.O.M.M. volunteers activated from shelters guided by HAM operators with emergency generators. Deploying boats into howling voids, C.O.M.M. braved fierce seas, navigating debris-cluttered routes and reaching anyone callers reported trapped. Radio contacts coordinated, avoiding duplicate rescues or overlooked dwellings. As official assets arrived once winds dropped, C.O.M.M. integrated bolstering efforts.

Aiding the Medically Fragile

Hazel, wheelchair-bound with chronic illness requiring oxygen, sheltered alone after refusing evacuation. When pumps failed amid power outages, neighbors alerted C.O.M.M channels. Volunteers Nicole and Chad immediately loaded a portable generator plus backup tanks into their boat, fighting treacherous swells. Reaching Hazel's isolated neighborhood, they located her submerged house through volunteers ashore guiding via radio. Nicole ascended the chimney carrying supplies while Chad steadied the boat against ferocious wind and current. Within hours, Hazel was stabilized in an ADA shelter with power-sustaining equipment. Had volunteers not circumvented breakdowns, her condition risked catastrophic deterioration alone.

Lessons in Community Preparedness

This real scenario emphasizes how citizen-led efforts empower prepared communities to stand together cooperatively. When conventional aid falters, radios allow distributed emergency networks to fill care gaps proactively. Inclusive training helps tailor plans for disabilities and language barriers, benefiting all. Solutions require ingraining resilience across social strata before disaster strikes rather than reactionary scrambling after impacts. Radio organizers proved grassroots coordination, strengthening evacuations amid infrastructure collapses and isolating those most vulnerable from life-saving support systems. Their story inspires community-scale preparedness, mitigating suffering through solidarity and self-sufficiency.

Case Study 3: Covert Surveillance of Illegal Logging Activity

Vast tracts of pristine woodlands remain vulnerable to exploitation across the developing world. When criminal elements threaten fragile ecosystems, concerned citizens sometimes resort to stealthily documenting offenses, enabling due process. This account portrays how radio operators discretely tracked poachers, jeopardizing an indigenous reserve, securing evidence, and allowing authorities to handle the situation.

Monitoring Suspicious Vehicles in the Reserve

Juan patrolled radio frequencies as a forest guard deep in the Amazon, vigilant for intruders endangering biodiversity. One dawn, he noticed traffic on reserved channels concerning two pickups entering the Parú Reserve under cover of mist, an irregular route. Switching to a covert channel, Juan discreetly followed trucks' rumbling through dense foliage using directional antennas. Intent on establishing poaching patterns without detection, Juan maintained radio silence, letting trucks range undisturbed but carefully marked routes via GPS. From treetop vantages, he noted crews halting felling valuable mahogany, cedar, and virola despite regulations. Returning at nightfall along new roads gouged into fragile soils, the poachers spoke arrogantly on open channels, emboldened by isolation from oversight. Juan realized curbing such wholesale destruction required discreet planning coordinated across many eyes and ears.

Enlisting a Covert Network of Observers

Discreetly contacting park collaborators, Juan cautioned against any transmissions risking detection by poachers' scanners. Together, they discreetly amassed equipment: HAM radios capable of encrypted Secure Digital modes, solar battery packs, infrared camera drones, and GPS units. Training in stealth observation and image transmission protocols, volunteers covertly deployed teams to monitor logging hotspots. Around the Parú, discreet eyes tracked plunderers' camps rising like sores across folded ridges. Operatives used low-visibility modes conveying haul sizes and routes plied by massive caterpillar tractors flattening broad swaths. Synthesizing such situational awareness across a distributed network allowed assessing logging's full ecological toll while evading poachers' detection or interference with legal protections.

Amassing Evidence Warranting Interception

Over months, the covert network methodically documented devastating impacts: over 500 hectares razed, countless animals butchered for bushmeat. They precisely located clandestine airstrips transporting pillaged treasures beyond borders for blood money. Recording license plates and crews' faces, the evidence irrefutably implicated complicit officials. When trucks gathered for a mass extraction, the network discreetly guided park rangers intercepting along hidden trails. Thanks to disciplined coordination between protectors adept in stealth and forensics, justice curtailed criminal schemes in a victory, restoring hope for biodiversity and emboldening additional guardians. Their story inspires grassroots efforts to protect vulnerable habitats through nonviolence, cooperation, and gadgets wielded responsibly.

Lessons in Distributed Vigilance

This scenario highlights how radios are skillfully applied to uphold sustainability and ethics and prevent escalation when authority breaks down. By circumventing detection, a dispersed network patiently amassed irrefutable intel, avoiding provocation and enabling proper process. Their triumph stemmed from discipline, forensics, and restraint over force, a model of benevolent defiance preserving wilderness through cooperation and law. For anyone striving conscientiously against wrongs too vast for an isolated response, radios offer empowering equalization, aiding justice through nonviolence and facts. This tale inspires citizen stewardship, safeguarding our natural heritage.

Essential Radio Frequencies for Emergencies

In any crisis scenario, the ability to call for help can be critical. However, conventional communication methods are sometimes unreliable during emergencies due to infrastructure damage or overloaded networks. Radios provide an off-grid option for summoning aid using public frequencies that are constantly monitored.

146.520 MHz - National Interoperability Channel

Known as the National Calling Channel, 146.520 MHz is monitored across the U.S. by various emergency response teams, including police, fire, EMS, and some government agencies. It provides a common channel for services to coordinate multi-agency responses. During large-scale emergencies like natural disasters, this frequency allows local departments to quickly request personnel or equipment from neighboring regions if needed. It can also be useful for civilians lost or injured in remote areas to attempt contact with any patrolling units.

155.475 MHz - State Interoperability Channel

Similar to the national channel but focused on a state level, 155.475 MHz is predominately monitored by public safety teams within individual states. This provides a local alternative to the national frequency if needing assistance confined within state borders. It may connect with emergency managers, search and rescue, national guard units, or highway patrols better positioned for response times compared to out-of-state agencies. During regional disasters isolating communities, this channel empowers relaying needs to state-based assets.

462.5500 MHz - National Simplex Interoperability Channel

As an interoperability option without a repeater, 462.5500 MHz allows simplex or "talk-around" communication between any two radios within range of each other. This can facilitate short-term on-scene coordination between the first responders arriving at an accident scene or tactical teams combing a disaster impact zone where infrastructure is down. It provides backups when repeaters used on other channels temporarily lose function. As an open communication line, passing safety alerts person-to-person directly benefits any encountered.

156.7 MHz - Distress & Safety Channel (VHF-FM Marine Band)

Monitored around the clock by Coast Guard stations, 156.7 MHz is the primary channel for vessel crews in coastal waters to declare emergencies. While not designed for land-based traffic, this frequency works well near shorelines and waterways, potentially connecting boaters, swimmers, or those isolated by flooding with Coast Guard air-sea rescue assets. Even inland, continued monitoring provides ocean/river community support. During tsunamis or hurricanes causing widespread evacuation, relaying distress here links coordinated maritime evacuation transport.

NOAA Weather Radio Stations (VHF public bands)

Broadcast across several VHF frequencies depending on region, NOAA Weather Radio transmits continuous 24/7 weather and emergency alerts directly from National Weather Service offices. Receivers can receive hazard warnings like tornadoes, hurricanes, or flash flooding, potentially saving lives with advanced notice. The signal also identifies the transmitter location, aiding rescuers following distress signals homing in on frequency. For updates on disasters disrupting other media, these stations disseminate real-time watches and warnings outdoors or from vehicles/boats.

CHAPTER 08

Troubleshooting and Maintenance

Common Issues and Solutions

As with any electronic device, Baofeng radios can sometimes experience technical issues that require troubleshooting and maintenance to resolve. Keeping your radio in good working order will ensure you always have a reliable means of communication.

Weak or Intermittent Audio

One of the most frequent complaints regarding Baofeng radios is weak or intermittent audio coming from the speaker. There are a few potential causes for this:

Dirty Volume or Squelch Controls: Over time, skin oils and dirt build up on the volume knob, squelch knob, and surrounding controls can make them sticky and less responsive. This decrease in conductivity between the contacts is a common culprit behind weak, crackling audio. The fix is simple - clean the knobs and surrounding casing thoroughly with electronic cleaner or isopropyl alcohol on a cotton swab or ball of paper towel. This removes contaminants and allows smooth, unrestricted movement again.

Faulty Speaker: On older radios, the actual speaker component can become damaged or detached from overuse. Test this by placing a clean cotton swab against the speaker grill and slowly turning up the volume. If no sound is heard, the speaker likely needs replacement. Baofengs uses standard 16-ohm speakers available for a few dollars online. Replacement only requires basic soldering skills.

Intermittent Wiring: Loose or damaged wiring inside the radio housing where the speaker connects can cause intermittent or completely absent audio. Gently flexing wiring while testing volume may reveal a loose connection that requires soldering or heat shrink tubing to secure. As a precaution, consider heat-shrinking all internal connections during routine maintenance.

Failure to Transmit

Another frustrating problem is when the radio appears to receive signals fine but is unable to transmit, as shown by the TX light not illuminating when the PTT button is pressed. Common transmit issues and their fixes include:

Dirty Mic Jack: The microphone jack, where the optional external mic connects, can get dirty inside from use. Test by switching to the internal mic - if the transmit works, the external mic jack needs cleaning. Use electronic cleaner applied to a cotton swab or paper clip to remove residue from the pin contacts.

Weak Mic Switch: Over time, the PTT switch button may lose conductivity and responsiveness. Gently clean around the switch with electronic cleaner without getting any liquid itself. The switch may need very light filing with fine-grit sandpaper to remove oxidation if cleaning doesn't help.

Faulty Wiring: As with audio issues, loose or broken internal wires where the mic connects inside can interrupt the transmit circuit. Carefully inspect for damage, reflow solder joints, or add heat shrink as needed. Consider upgrading to an external heavy-duty PTT mic for long-term durability.

Poor or No Reception

Reception issues like inconsistent signal strength, intermittent drops in volume, or complete inability to pick up local transmissions have several potential root causes:

Dirty Controls: Volume, squelch, and tuner knobs may need a thorough cleaning, inside and out, to remove contaminants impairing smooth operation, as mentioned earlier.

Faulty Antenna: An SWR meter can verify if the antenna is still making an efficient radio-frequency match for transmitting and receiving. Replacements only cost a few dollars if necessary. One fix is that people are occasionally re-soldering the center-pin connection inside the antenna connector.

Loose Connection: Inspect where the antenna screws into the radio BNC connector for tightness and condition of the collar seal. A loose connection will impair signals getting in and out of the radio.

Low Battery: A weak battery may not be able to provide consistent power regulation to the radio circuits. Keep replacement batteries fully charged.

Programming/Channel Issues

Programming mishaps or faulty memories can also undermine the intended operation of the radio:

Failed Memory Storage: Static electricity or glitches can sometimes corrupt the channel memory, causing presets to get "lost." Reprogramming all channels may fix it.

Mis-programmed Channel: If a specific preset channel is unable to be recalled, there may be an issue with how that channel was originally stored, requiring reprogramming just that channel.

Stuck Buttons: Cleaning sticky key buttons with electronic cleaners can restore functionality if they seem "stuck" and unresponsive during programming or channel changes.

As with other issues, try cleaning programming buttons, flexing wiring near them, and reflowing nearby solder connections to rule out mechanical or electrical faults. Reprogram all channels as a troubleshooting step.

Intermittent Power or Switching Issues

On/off and power problems are frustrating as they can make the radio seem unreliable or broken when really minor fixes may resolve:

Dirty Power Switch: The on/off switch can build up grime, causing intermittent or weak contact. Gently clean the switch with an electronic cleaner without getting any liquid inside the radio.

Low Battery/Poor Contact: Weak batteries may power the radio for a while but cause it to suddenly switch off. Clean battery terminals and ensure a tight metal-to-metal connection. Consider higher-capacity replacement batteries.

Faulty Wiring: Inspect the wiring behind the power switch and run to the battery contacts for fractures or loose connections that can cause intermittent power drain. Reflow solder joints if needed.

As many issues stem from minor electrical faults, always start troubleshooting with cleaning controls, jacks, and connectors thoroughly before diving deeper into a radio. Replacing batteries regularly also prevents unexpected glitches.

Cleaning and Protecting Your Baofeng Radio

Exterior Cleaning

The exterior sees the most grime accumulation, so it should be wiped down frequently. Dampen a microfiber cloth with distilled water only - avoid harsh chemicals that could damage plastics. Gently wipe all casing areas, crevices, and control surfaces to remove skin oils and dirt buildup. Pay special attention to controls and ports that see contaminants from frequent touching and cable insertion. Let fully dry before use to prevent water from infiltrating seams or conducting into the radio internals. Consider an electronics-safe spray like compressed air for hard-to-reach areas like Speaker grilles.

Interior Cleaning

Periodically accessing the inside removes even more deposits. Remove rear screws/covers carefully per the user manual and anti-tamper seals if present. Compressed air blows out dust from crevices where fingers can't reach, like antenna mount threads. Use electronics cleaner on a dry cotton swab to wipe PCB boards and visible metal contacts. Replace desiccant packs if present to maximize moisture protection during storage. Check wires for damage, cold solder joints, or breaks before reassembly.

Connector Hygiene

Frequent contaminant culprits need regular, dedicated attention. Mic/speaker jacks collect debris, affecting audio quality over time. Twist a dry cotton swab or coffee filter inside jacks while wiping to clean all surfaces. Apply isopropyl alcohol to external jack openings with Q-tips for a deeper clean. Clean power and antenna contacts similarly suffer from repetitive connections.

Protective Equipment

Proper storage shields radios from unnecessary exposure when not in use. Hard cases protect from dings, moisture, dust/dirt during transport and storage. Soft pouches shield displays and controls without excess rigid casing. Lanyards prevent radios from being misplaced when not in a case. Desktop stands to keep screens clean and controls accessible in a workspace. Weatherproofing accessories like antenna-mounted caps safeguard during weather events.

Proper Storage Practices to Prolong Radio Life

A Baofeng radio is a valuable tool for both work and recreation, but it requires careful handling and storage when not in use to maximize its lifespan. Without proper storage practices, accumulated wear can accelerate and potentially lead to costly repairs.

Protective Cases

Using a durable carrying case is essential for any radio seeing intermittent usage. Hard shell cases prevent impact damage like cracks or dents that could occur from accidental drops while traveling with the radio. They also provide protection from dirt, dust, and moisture ingress that can degrade internal components over time. Cases with sealed closures, soft interior lining, and closed-cell foam padding provide optimal protection against these contaminants while cushioning the radio from abrasive surfaces. Air-tight zippers further ensure no particulates make their way inside the stored radio.

Environmental Controls

Prolonged exposure to temperature or humidity extremes can negatively impact component reliability far into the future. It is recommended to regulate storage environments between 40-80°F and 40-60% relative humidity to avoid issues from heat, cold, or condensation. Silica gel packs inside protective cases help absorb moisture, which can accelerate corrosion. Storing away from chemical fumes or cleaning supplies protects the radio from potential airborne pollutants as well.

Handling & Inspection

Mistreatment during dormancy periods can shorten a radio's useful service life unnecessarily. Excessive flexing of ports or button pressing should be avoided to protect delicate internal connections. Even slight impact drops concentrating stresses in vulnerable areas increase risks of cracks forming over many usage and storage cycles. Periodic inspection of casing seals and integrity every six months alerts owners to any needed repair or replacement before potential damage escalates.

Conductive Contacts

Oxidation left on the battery and port thread contacts poses risks to proper functionality upon future use. Thorough cleaning and application of electrical contact preservatives like DeoxIT protect vulnerable connection points relied upon for powering the radio. Anti-corrosion grease on battery terminals prevents electrolyte cell formation from ambient moisture exposure during long-term storage periods.

Handling Extreme Weather Conditions and Environmental Challenges

While Baofeng radios offer value, their construction is not as rugged as high-end outdoor models. However, with preparation and care, these radios can still function reliably in challenging environments.

Heat Exposure

Excessive heat quickly degrades components if not mitigated properly. It is important to store radios sealed in vehicles or equipment to regulate temperature during very hot conditions. Allow adequate airflow and protect from direct sun exposure to prevent internal overheating. Check radios frequently when temperatures exceed 100°F for potential issues. Taking breaks from heat and preventing enclosed overheating prevents thermal damage.

Cold Environments

Cold temperatures inhibit battery power and impact display readability. Shield-stored radios with insulated cases to retain warmth in frigid vehicles. Gently warm radios before use, near or below freezing, to avoid condensation. Monitor lithium-ion battery capacity loss in the cold and change as needed. Wearing insulating gloves allows for operating cold-numbed controls comfortably. Proper temperature management defeats cold-related issues.

Moisture Dangers

Exposure to water leads to corrosion, so protection and swift drying is crucial. During rains or wet work, radios must be under shelter or sealed cases. Removing batteries and using drying packets helps purge lingering moisture promptly after exposure. Inspect for water intrusion signs and disassemble to dry if needed. Applying conformal coating or tape adds moisture resistance in harsh areas.

Dust and Particulates

Avoiding dust and particle blockage preserves cooling and connections. Frequent compressed air cleans crevices and buttons. Use vented cases or dust caps on ports except during direct use. Attach filters to external antennae to trap debris better than internal protections alone. Consistent cleaning extends equipment life in difficult environments.

Recommendations and Preventative Maintenance

Beyond reactive troubleshooting, some proactive steps and supplementary gear purchases can maximize a Baofeng radio's dependability and functionality over the long haul:

- Extra Batteries - Have backup sets of high-capacity rechargeable batteries on hand. Rotate charging old batteries to keep them fresh.
- Routine Cleaning - Wipe down controls periodically with the electronic cleaner to prevent buildup. Do a deep interior clean once a year.
- Store in Case - Keep radios in protective carrying cases when not in use to prevent grit and debris from accumulating on contacts.
- External Antenna - For long-range communication, an efficient Nagoya or Diamond antenna typically nets much better sensitivity than stock stubs. Experiment to find the right style.
- Signal Monitor - An SWR meter lets you routinely check antennas are tuned for the most efficient radiation and reception. Replace if detuned.
- Heavy Duty Mic - Swapping the flimsy stock mic for a sturdier aftermarket mic prevents potential failures from everyday use.

- Drop-In Rapid Charger - Saves having to charge batteries externally one at a time on a slow wall wart.

By addressing problems early with diligent cleaning and minor maintenance and upgrading weak points selectively, Baofeng radios can certainly last for years of hobby and public service use while rarely leaving the owner high and dry. With some basic troubleshooting know-how, owners can keep their radios performing flawlessly for all two-way communication needs.

CHAPTER 09

Regulatory Compliance

FCC Regulations

As any technology relying on radio spectrum for transmission of voice, data, and other signals, amateur radios like the Baofeng are regulated in the United States by the Federal Communications Commission (FCC). This chapter will explore the FCC rules and regulations amateur operators must follow to stay compliant with the law. Understanding these guidelines is important for the legal and safe use of your radio.

The FCC and its Authority

The FCC is an independent regulatory agency of the United States government that oversees interstate and international communications via radio, television, wire, satellite, and cable. It was established by the Communications Act of 1934 and has jurisdiction across these various media types. The FCC aims to make communication services available to all Americans while balancing priorities like availability, affordability, competition, national defense, public safety, and efficient use of the spectrum. They promote the overall public interest through rules that allow maximum benefit from limited radio frequencies. All wireless devices operating or transmitting within the U.S, including amateur radios, fall under FCC authority. The agency works to manage the shared public airwaves for the coexistence of important services like radio/TV broadcasting, wireless technology, emergency response, amateur radio, and citizen bands. Proper enforcement of regulations helps prevent signal interference issues.

FCC Regulations for Amateur Radio

Part 97 Rules

The FCC has established a section of rules specific to the Amateur Radio Service known as Part 97. These dictate proper operating procedures, authorized frequencies, power limits, and other technical specifications that licensees must adhere to. Some key aspects covered in Part 97 include:

- Permitted communications are related to self-training, intercommunication, and technical studies only. Commercial use is prohibited.
- Transmissions cannot be encrypted or contain any music, advertisements, obscene/indecent content, or communications intended to facilitate criminal plans/acts.
- Stations must transmit the assigned call sign at least every 10 minutes during contacts and at the session conclusion. This supports proper identification.
- Operators are restricted to amateur radio frequency allocations designated by the FCC and cannot exceed power limits based on license class (Technician, General, or Extra).
- Transmitter power must be set to the minimum required for successful communication and must never exceed the legal limit.

Matters like international operations, special temporary authorities, and specific technical standards are also covered under Part 97 regulations.

License Requirements

While receiving broadcasts is unregulated, transmitting on amateur frequencies via a Baofeng or any radio requires a valid FCC-issued operator license. Three classes exist - Technician, General, and Extra - each allowing different operating privileges corresponding to demonstrated technical knowledge. Proper testing ensures operators learn sound operating practices before getting on the air.

Operator Licenses

To thoroughly explore FCC rules, we will delve into more detail about amateur radio licenses:

Technician Class License

The entry-level Technician license is earned by passing a 35-question test over basic radio law, practices, and theory. It grants privileges on VHF/UHF frequencies above 30MHz, like the very popular 2-meter and 70cm bands. Maximum output power is limited to 50 watts.

General Class License

A more advanced 50-question exam must be completed to obtain General class privileges. This adds HF/shortwave allocations from 1.8-54MHz with a higher power limit of 150 watts. It demonstrates proficiency in radio regulations and principles.

Extra Class License

The top-level Extra Class operator has passed the most difficult 35-question exam focused on advanced concepts. It confers all frequency privileges and a power cap of 1,500 watts. Extras showcase extensive expertise in the technical facets of amateur radio operations.

Operator responsibilities

Obtaining any FCC license comes with ongoing requirements. Operators must follow Part 97 rules, use identification procedures like call signs properly, avoid harmful interference to other users, and not exceed their license privileges. Continuing to expand knowledge over time is also encouraged.

FCC Enforcement

While promoting a collaborative regulatory approach, the FCC takes violations seriously to maintain a fair operating environment. Penalties range from warnings for minor infractions to large fines, license suspensions, or imprisonment, depending on the nature and repetition of misconduct. Cooperation is the wisest response if enforcement action occurs to potentially reduce consequences.

Understanding and Adhering to FCC Rules

The Federal Communications Commission, or FCC, is the government agency responsible for regulating all wireless communications in the United States, including amateur radio. As a licensed amateur radio operator, understanding and adhering to FCC rules is crucial. Non-compliance can result in penalties, and maintaining orderly spectrum usage benefits all users.

The Federal Communications Commission

The FCC was established by the Communications Act of 1934 and acts independently of other government branches. Its core mission is to make radio, television, wire, satellite, and cable communications services available to all Americans while balancing priorities around availability, competition, and innovation. The FCC oversees technology authorization, assigns spectrum bands, enforces statutes and regulations, and addresses complaints to remain in the public interest. All amateur radio stations must recognize the FCC's authority in these areas.

Rules for the Amateur Radio Service

The FCC regulations specific to amateur radio are outlined in Part 97 of the Code of Federal Regulations. Key aspects covered include permitted communications like self-training and technical experimentation, identification procedures, frequency allocations, power limits, station control responsibilities, and modulation standards. Part 97 also contains enforcement provisions for violations. Understanding and adhering to these rules maintains operating privileges and avoids sanctions. The FCC takes a balanced yet serious approach to compliance.

Operator Licensing Requirements

While receive-only equipment requires no license, transmitting necessitates FCC authorization to demonstrate knowledge. Three classes exist, offering increasing privileges based on examination—Technician, General, and Extra. Licensing ensures operators learn sound practices like regulations and safe operation before getting on the air. Licenses must remain current to retain eligibility and compliance.

Technical Operating Standards

Comprehensive knowledge includes FCC technical standards. Key areas are proper equipment certification, emissions contained within limits, control operator supervision of authorized transmissions, and permitted modulation types.

Attention to technical specifications satisfies FCC equipment and parameter guidelines.

Responsible Operating Practices

The FCC promotes cooperation to address issues when possible. However, blatant or harmful violations face enforcement actions and potential penalties. Licensees carry ongoing responsibilities like courteous communication, addressing interference promptly, cooperating fully with FCC inquiries, and understanding willful misconduct can incur severe consequences. Voluntary compliance supports amateur radio's continued role.

Best Practices for Responsible and Ethical Radio Communication

As an amateur radio operator, it is important to conduct all transmissions in a responsible and ethical manner. Proper operating procedures help ensure effective communication while avoiding interference or other issues.

Calling Procedures

Proper calling procedures allow stations to make contact without interfering with ongoing exchanges. Typical steps include listening first to avoid duplicating calls, waiting your turn if a frequency is busy, identifying your station clearly and completely, and keeping transmissions brief. This respects other users' operating time.

Courteous Conversation

During QSOs, courteous operators communicate politely and avoid topics unacceptable in public discourse like politics, religion, or offensive subjects. Volume is kept at a reasonable level without yelling or screaming into the microphone. Lengthy socializing should make way for others waiting to use the frequency.

Interference Avoidance

All radios have some potential to cause interference, so best practices minimize this risk. Adjusting transmission power to the minimum effective level, properly maintaining equipment, and monitoring frequencies before transmitting help prevent disruption of other communications. Operators quickly resolve issues if alerts occur.

Secure Communications

Most amateur radio communications are open and observable to anyone listening. While some digital modes allow encryption, the FCC requires prior approval for such transmissions. Operators do not use radio systems to facilitate illegal plans or activities between parties.

Emergency Responsibilities

When normal channels become busy during emergencies, seasoned radio amateurs switch to alternate stations trained in emergency response disciplines. Communication of safety hazards or requests that compromise responder efforts should be avoided unless part of an official incident response role.

Complaint Resolution

Should interference problems occur, responsible operators make every effort to promptly troubleshoot the issue and resolve it in a practical, professional manner. Escalating to FCC enforcement action is a last resort if direct resolution attempts fail. Cooperating with any FCC inquiries is expected.

Continuous Improvement

Advancing one's radio skills through prudent technical experimentation, specialized training programs, and offering assistance to newcomers benefits the entire amateur community. Regular assessments identify potential areas for individual licensees to strengthen compliance and on-air conduct over time.

CONCLUSION

In this book, we have covered the essential information needed to master your Baofeng radio and effectively use it in a variety of survival and emergency situations. From the basics of how the radio works and its features all the way to advanced communication techniques and real-life case studies, the goal has been to equip you with the knowledge and skills to enhance your communication abilities when standard infrastructure fails. As with any tool, it is important to understand both how to properly operate your radio as well as the legal requirements around its usage. In this conclusion, we will summarize the key lessons and reiterate some best practices to help ensure you get the most value out of your Baofeng radio.

One of the primary reasons for choosing a Baofeng radio is its versatility and affordability relative to other brands. As covered in the first chapter, the UV-5R model, in particular, offers dual-band capability on both VHF and UHF frequencies, a wide receiving range, programming options, an extensive aftermarket for accessories, and a sturdy build quality - all at a very reasonable price point. However, to fully take advantage of these features requires knowing how to program your specific channels, settings, and other options either manually or through CHIRP software. Spending time upfront to customize your radio optimizes it for your specific needs and locations. Always keep backup documentation of your programming just in case you need to reset or reprogram the radio in an emergency. As evidenced by the case studies in Chapter 7, having a radio like the Baofeng UV-5R can truly make a lifesaving difference in a variety of survival and tactical scenarios. From responding to hurricanes and wildfires to search and rescue missions or protective monitoring of public demonstrations, these affordable yet versatile radios have time and again helped first responders and community groups maintain command and control when other options fail. Their value is further magnified when coupled with the knowledge gained through comprehensive guides like this one. Study the real examples closely to help envision how you could employ similar strategies in your own emergency response plans or survival retreat operations.

Of course, to maintain legal compliance, it's important to understand the FCC regulations regarding which frequencies your radio can access based on your license class. For emergency or tactical usage without a ham license, carefully selecting FRS/GMRS channels is recommended over broad VHF/UHF monitoring. Nonetheless, if procured and operated legally, the Baofeng UV-5R can fill an important niche as a capable yet low-cost communications tool. Also, pay attention to routine maintenance, like keeping batteries charged or swapping them regularly in storage. Consider spare parts, charging solutions, and magnetic antenna mounts as other auxiliary equipment multiplies the handheld's functionality.

In closing, a durable, programmable, multi-band radio like the Baofeng UV-5R empowers you to tackle communications challenges across a wide range of survival and self-reliance scenarios. However, its full potential can only be realized when paired with the human elements of planning, training, and established protocols. Use this guide as a starting point for expanding your skills in radio operation, emergency coordination techniques, and leveraging technology to strengthen your preparedness infrastructure. With a commitment to legal compliance and ongoing practice, a Baofeng in your toolkit readies you to aid yourself and others through times when that little voice becomes the most vital link. Stay sharp and stay safe.

Made in United States
Troutdale, OR
03/09/2024

18343382R00060